SpringerBriefs in Microbiology

Extremophilic Bacteria

Series editors

Sonia M. Tiquia-Arashiro, Dearborn, MI, USA
Melanie Mormile, Rolla, MO, USA

More information about this series at http://www.springer.com/series/11917

Lesley-Ann Giddings · David J. Newman

Bioactive Compounds from Extremophiles

Genomic Studies, Biosynthetic Gene Clusters, and New Dereplication Methods

 Springer

Lesley-Ann Giddings
McCardell Bicentennial Hall 447
Middlebury College
Middlebury, VT
USA

David J. Newman
Natural Products Branch,
    Developmental Therapeutics Program
NCI
Frederick, MD
USA

ISSN 2191-5385   ISSN 2191-5393   (electronic)
SpringerBriefs in Microbiology
ISBN 978-3-319-14835-9   ISBN 978-3-319-14836-6   (eBook)
DOI 10.1007/978-3-319-14836-6

Library of Congress Control Number: 2014959120

Springer Cham Heidelberg New York Dordrecht London

Printed on acid-free paper

Springer International Publishing AG Switzerland is part of Springer Science+Business Media
(www.springer.com)

Contents

Bioactive Compounds from Extremophiles

Abstract Microorganisms continue to serve as valuable sources of natural product therapeutics for the treatment of disease, evidenced by roughly 50 % of approved small molecule antitumor drugs being either directly from or inspired by microbes. The continuous number of drugs derived from these agents that are approved each year remains unmatched by compounds synthesized via combinatorial chemistry, renewing scientific interest in natural product drug discovery. Because researchers exhausted the "low-hanging fruit" produced by microorganisms inhabiting terrestrial environments in the early 1990s, the search for new sources of bioactive agents has now expanded to extremophiles found in unique terrestrial and marine environments. Here, we define extremophiles as single-celled microorganisms that inhabit non-mesophilic environments characterized by high salinity, extreme temperatures and pressures, and variable acidity. Importantly, major breakthroughs have been made in the fields of molecular biology, genomics, bioinformatics, data mining, chemistry, and biochemistry, resulting in rapid, new integrative ways of identifying novel natural products from previously unculturable microorganisms or complex extracts prepared from both mesophiles and extremophiles. The authors review how next-generation sequencing platforms have speed up the discovery of bioactive metabolites and provided a blueprint of the genetic capability of microorganisms to produce new pharmacophores. New methods for activating cryptic gene clusters as well as heterologously expressing genes with unknown functions are described. In addition, new analytical instrumentation, dereplication and statistical analysis methods, and their integration with genomic sequence data used for the isolation and structural elucidation of natural products are discussed.

Keywords Next-generation sequencing technology · Heterologous gene expression · Metagenomics · Cocultivation · Extremophiles · Dereplication methods · Natural products · Microorganisms · Liquid chromatography · Mass spectrometry · Nuclear magnetic resonance · Principle component analysis · Hierarchical cluster analysis

© The Author(s) 2015
L.-A. Giddings and D.J. Newman, *Bioactive Compounds from Extremophiles*,
Extremophilic Bacteria, DOI 10.1007/978-3-319-14836-6_1

1 Introduction

Nature's chemical arsenal is composed of unique chemical scaffolds with diverse functional groups and stereochemistry that interact with various biological targets within or outside a host. Natural product chemists have been exploiting these biological activities over the last 70 years or so to develop new therapeutics via target-based screening programs and understand the modes of actions of drugs. Historically, the protocol for microbial natural product discovery has involved the isolation of microorganisms, fermentation of individual strains under standard fermentation conditions, and extraction of culture broths with organic solvent. If an extract exhibited any activity in a bioassay, then each compound within the mixture was isolated and their biological activity was evaluated. Furthermore, in order to maximize the potential for a microorganism to produce a unique and abundant metabolome, fermentation conditions, such as the pH, temperature, growth time, and media, were commonly modified. Using these methods, chemists isolated "low-hanging fruit," molecules produced by terrestrial microorganisms that could be easily grown under standard laboratory conditions. However, due to the high rediscovery rates of compounds, many pharmaceutical companies de-emphasized their natural products discovery programs in the 1990s, as new microbial sources were needed.

Within the last 20–30 years, researchers have started to shift their focus to isolating extremophiles with novel metabolomes from unexplored environments. For the purpose of this book, we define extremophiles as single-celled microorganisms (e.g., archaea, eubacteria, and fungi) living under non-mesophilic conditions, such as in environments with high salinity, high radiation, limited nutrients, extreme temperatures and pressures, variable pH, or any combination thereof. These extremophiles can be found on land as well as in the marine world, which represents some of the world's most extreme environments. The slow but continuous outflow of publications on the isolation and bioactivity of new secondary metabolites from extremophiles are a result of major challenges researchers still encounter, such as recreating the conditions of the microorganism's native environment for fermentation purposes, identifying and isolating the true microbial producer from a consortia of organisms, activating the expression of biosynthetic gene clusters that are not expressed under standard laboratory conditions, developing high-throughput screens to identify new potential drug targets, and obtaining enough pure compound for structural elucidation and (pre)clinical studies. Recently, more time and resources have been invested in expanding the toolbox of chemists biochemists, pharmacologists, microbiologists, and molecular biologists to overcome these bottlenecks in drug discovery, as over 50 % of approved anti-infective and antitumor agents are naturally derived or inspired (as of 2010) (Cragg and Newman 2013). Roughly 50 % of approved small molecule antitumor drugs are directly from or inspired by microbes as of 2012 (Giddings and Newman 2013).

With the surge in the number of sequenced microbial genomes, bioinformatic analyses of these genomes have revealed that scientists have greatly underestimated the genetic capabilities of even the most well-studied microbes to produce a vast array of compounds, suggesting that most of the biosynthetic gene clusters involved in microbial secondary metabolism are cryptic or expressed at low levels. For example, sequencing *Streptomyces* genomes (*S. avermitilis* and *S. griseus*) as well as *Saccharopolyspora erythraea* demonstrated that each strain has the potential to produce at least 20 or more secondary metabolites, while only a small subset of these are produced by conventional fermentation methods (Ohnishi et al. 2008). The genetic capability of fungi has also been considerably underestimated as the model soil fungus *Aspergillus nidulans* has been reported to have 28 putative polyketide synthase (PKS) and 24 non-ribosomal peptide synthetase (NRPS) gene clusters, demonstrating its capability of producing at least 52 secondary metabolites (von Döhren 2009). In their 2012 review, Sanchez and coworkers have also summarized the number of PKS, PKS-like, NRPS, and NRPS-like genes in several sequenced *Aspergillus* genomes, further highlighting the potential of fungi to produce novel natural products (Sanchez et al. 2012). Considering that microbes from marine and other extremophilic environments remain to be fully exhausted due to limited access and difficulty in culturing them under standard laboratory conditions, these extremophiles represent an untapped source of bioactive secondary metabolites. Notably, only <0.1 % of marine-derived microbes have been isolated (Simon and Daniel 2009), leaving a ubiquitous number of unique metabolomes to be surveyed. Thus, genomic data have brought natural product producers back into the spotlight for reevaluation.

Now that we know that there is a greater potential for microorganisms to produce secondary metabolites, several advances have been made in the production and isolation of new compounds. New culturing techniques and culture-independent methods have been developed to activate the expression of cryptic gene clusters to identify new natural products. As the genes involved in secondary metabolism in microorganisms are typically clustered together in a genome, scientists have developed in silico methods to predict the structures of some of the secondary metabolites that are not produced under mesophilic fermentation conditions. This approach to finding natural products is referred to as genome mining, which involves exploiting sequence data to produce new natural products by activating transcriptionally silent gene clusters by overexpressing or downregulating the expression of positive or negative transcription regulators in the producing organisms or using heterologous hosts to enhance the production of cryptic metabolites. This book examines the culture-dependent and culture-independent techniques currently used to identify new novel metabolites as well as some of the new analytical methods (e.g., LC-MS, LC-NMR, and PCA analysis) that have been developed for profiling and structurally characterizing large metabolomic datasets.

2 Activating the Expression of Natural Product Biosynthetic Gene Clusters

One of the major challenges in natural product drug discovery is the lack of an economical way to produce large quantities of compound for clinical evaluation, making secondary metabolites from microbial sources more appealing bioactive agents to work with because they can be replenished via fermentation once their corresponding biosynthetic genes are expressed. Traditionally, microbial hosts have been manipulated or reengineered to increase product titers (Gomez-Escribano et al. 2012), but this strategy can be labor-intensive and time-consuming, commonly involving several iterative rounds of chemical or irradiative mutagenesis followed by the screening of mutants for a desired phenotype. Even with significant optimization of the producing strain, biosynthetic gene clusters often are still expressed at low levels. Heterologous expression of biosynthetic gene clusters is one of the main strategies used to improve the expression of these genes as well as characterize and modify biosynthetic pathways to produce new natural products and derivatives, especially now that genomic studies have revealed the abundance of "silent" biosynthetic gene clusters in microbial genomes.

E. coli has been engineered to produce several secondary metabolites from actinomycetes (Rodriguez et al. 2009) and several other hosts, such as *Pseudomonas putida* and *Myxococcus xanthus*. However, the size limitations (~100 kb) associated with some of the gene products encoded by large gene clusters, such as polyketide synthetases (PKSs) or non-ribosomal peptide synthases (NRPSs), often make heterologous expression difficult or inapplicable. Other challenges encountered are the inability of the host to express correctly folded proteins and then to post-translationally modify via biosynthetic enzymes, as well as produce the required biosynthetic precursors. Scientists have circumvented these problems by manipulating the expression of pathway regulators (e.g., α-amylase promoters, inducible *alcA* promoters, and LuxR regulators) to induce the expression of cryptic genes, using specific vectors or hosts, and introducing or deleting competing biosynthetic pathways. New culture-based methods, such as modifying components of fermentation media and coculturing microorganisms, have also emerged as valuable techniques to activate the production of unknown metabolites.

2.1 Genetic and Metabolic Engineering

The significant advances made in recombinant DNA technology, such as expression vectors and site-directed mutagenesis, have provided tools for genetically engineering microbial cells to characterize and manipulate genes involved in the production of secondary metabolites. These tools combined with the recent advances made in genomic sequencing have enabled scientists to exploit genetic information to engineer microorganisms to produce new (un)natural products as well as optimize their titers. Novel (un)natural products can now be identified using user-friendly bioinformatics tools, such as antiSmASH (Blin et al. 2013), and their corresponding gene clusters can be assembled and modified. Importantly, not only can enzymatic products be predicted based on the substrate specificity of each domain, but now gene clusters from a variety of organisms can be mixed, matched, or mutated to produce new chemical entities.

Genetic engineering has been used to produce several bioactive, unnatural metabolites, such as the fluorometabolite fluorosalinosporamide A **1** and salinosporamide X7 **2** (Fig. 1) (Eustáquio and Moore 2008; Nett et al. 2009).

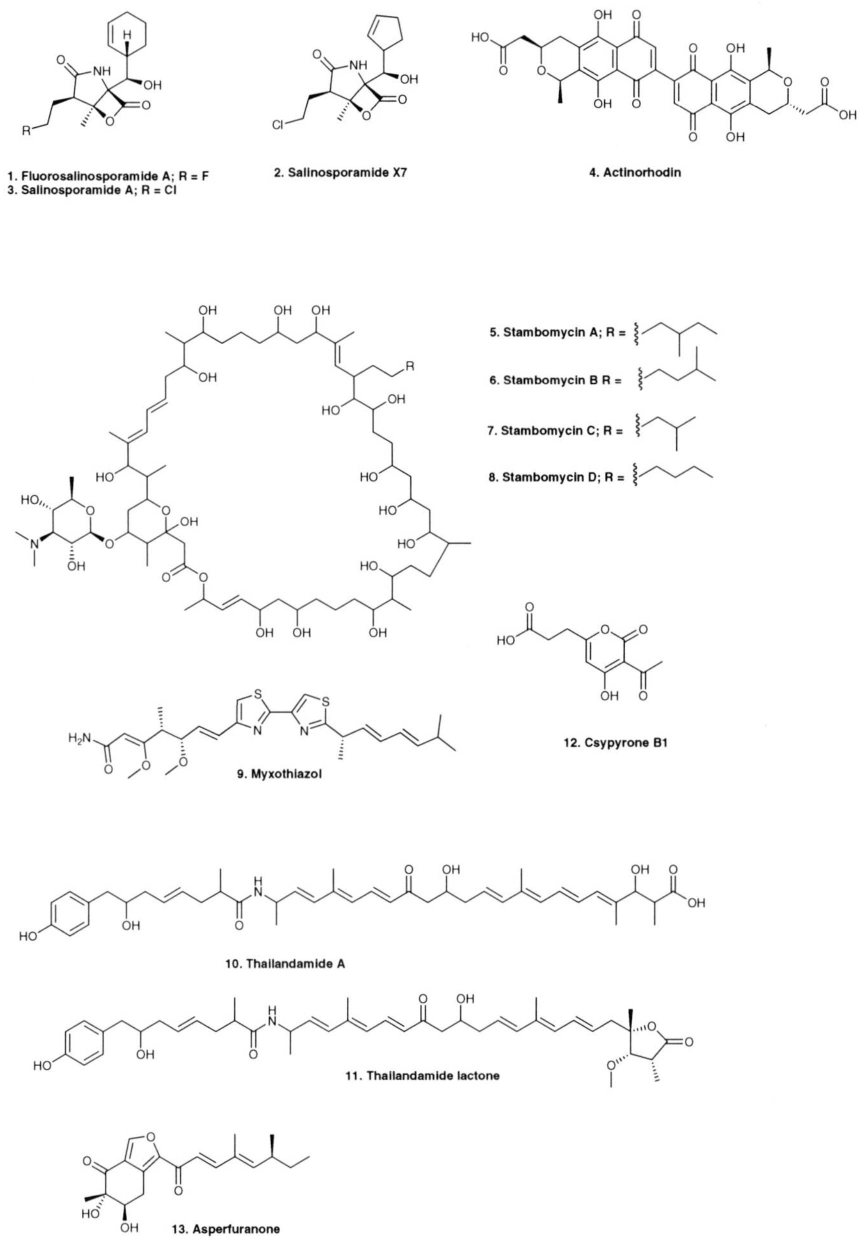

1. Fluorosalinosporamide A; R = F
3. Salinosporamide A; R = Cl

2. Salinosporamide X7

4. Actinorhodin

5. Stambomycin A; R =
6. Stambomycin B R =
7. Stambomycin C; R =
8. Stambomycin D; R =

12. Csypyrone B1

9. Myxothiazol

10. Thailandamide A

11. Thailandamide lactone

13. Asperfuranone

Fig. 1 Structures **1–13**

The Moore research group produced fluorosalinosporamide A **1** by silencing the chlorinase gene *salL* in *Salinispora tropica* followed by supplementing the fermentation media with synthetic 5-fluoro-5-deoxyribose, whereas salinosporamide X7 **2** was produced by inactivating the prephenate dehydratase homolog *salX* in *S. tropica* and supplementing the media with 3-*R*,*S*-cyclopent-2-enyl-D,L-alanine. These studies highlight how genetic engineering and precursor-directed biosynthesis can successfully be used to produce novel marine-derived analogs (Eustáquio and Moore 2008). Two years later, the same research group reengineered *S. tropica* to produce fluorosalinosporamide A **1** by replacing *salL* in *S. tropica* with the fluorinase gene *flA* from *S. cattleya* and growing the *salL⁻* *flA⁺* mutant strain in the presence of inorganic fluoride (Eustáquio et al. 2010). Fluorosalinosporamide A **1** was determined to be a slow reversible inhibitor of the 20S proteasome, whereas salinosporamide X7 **2** was nearly three times more cytotoxic than salinosporamide A **3** (Fig. 1). As over 150 fluorinated drugs have been FDA approved, comprising approximately 20 % of all pharmaceuticals, including the antidepressant fluoxetine (Prozac) and antibacterial ciprofloxacin (Ciprobay) (Müller et al. 2007), these studies are prime examples of how genetic engineering can be used to introduce unique functionalities in molecules.

In addition to producing novel chemical entities, such as fluorosalinosporamide **1**, microorganisms can also be engineered to optimize the titer of a desired product. Once a host cell, either the producing microorganism or another microorganism, is chosen, metabolic pathways can be added, deleted, or mutated to balance the metabolic flux within the cell and optimize the product yield (Ongley et al. 2013). The flux distribution of metabolites is often controlled by complex regulatory mechanisms and the expression of individual or multiple genes. Figure 2 summarizes some of the molecular methods that can be used to increase the expression of (cryptic) biosynthetic gene clusters. Mutating RNA polymerase, the

Molecular methods for optimizing native or heterologous expression of cryptic genes

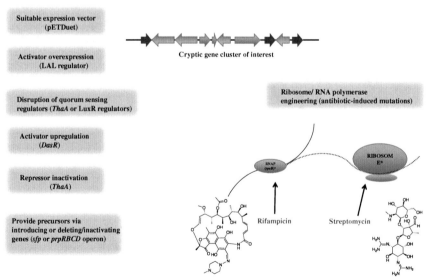

Fig. 2 Methods of increasing expression of "cryptic" biosynthetic clusters

ribosomal S12 protein, as well as other ribosomal proteins and translation factors can enhance the production of secondary metabolites. For example, the expression of cryptic gene clusters within *Streptomyces lividans*, which does not produce antibiotics, was improved by mutating endogenous ribosomal S12 protein to confer streptomycin resistance, resulting in the production of large quantities of the antibiotic actinorhodin **4** (Fig. 1) (Shima et al. 1996). The binding of alarmone guanosine tetraphosphate (ppGpp) to RNA polymerase has also been reported to trigger the production of antibiotics, influencing researchers to introduce rifampicin-resistant mutations into RNA polymerase, which mimic the ppGpp-bound form to increase the expression of cryptic gene clusters (Ochi 2007).

By mutating the genes encoding the β-subunit of RNA polymerase (*rpoB*) and ribosomal protein S12 (*rpsL*), Ochi and Hosaka were able to improve antibiotic expression in *Streptomyces* spp., myxobacteria, and fungi and produce the previously unknown piperidamycins (Hosaka et al. 2009; Ochi and Hosaka 2013). These modifications are not observed in wild-type microbial strains and have minimal effects on the native host. Furthermore, several actinomycetes have been reported to possess two *rpoB* genes that improve the overexpression of antibiotic biosynthetic genes. The use of mutated *rpoB*(S), rifampicin-sensitive wild-type *rpoB*; and *rpoB*(R), [rifampicin-resistant mutant-type *rpoB*] or duplicated *rpoB* paralogs provide two functionally distinct and developmentally regulated RNA polymerases that can activate or increase the expression of gene clusters in *Nonomuraea* sp. strain ATCC 39727 and *S. lividans* strains 1,326 and KO-421 (Talá et al. 2009).

Other targeted strategies have been used to enhance the production of secondary metabolites, such as overexpressing positive pathway-specific regulators, decreasing the expression of negative pathway-specific regulators, and disrupting quorum-sensing regulators, such as LuxR. For example, Laureti and coworkers overexpressed the pathway-specific activator LAL regulator, a large ATP-binding LuxR-type gene, under control of a strong constitutive promoter *ermE*p* in *S. ambofaciens*, which resulted in the expression of a silent, large type I modular PKS gene cluster involved in the production of four 51-membered glycosylated macrolides, stambomycins A–D (**5–8**) (Fig. 1) (Laureti et al. 2011). This cluster is composed of 25 genes and spans up to 150 kb, one of the largest PKS gene clusters reported. The stambomycins have unusual extender units into the polyketide chain, and *trans* hydroxylation is involved in the growing polyketide chain to incorporate the hydroxyl group required for macrolactonization. Notably, stambomycins A–D (**5–8**) exhibited antibacterial activity against Gram-positive eubacteria and antiproliferative activity against human adenocarcinoma HT29 cell lines with similar IC_{50} values ranging from 1.74 to 1.77 μM. Moreover, all compounds exhibited significant antiproliferative activity against human breast MCF7, lung H460, and prostate cancer PC3 cell lines.

Sandmann and coworkers also improved the yield of the highly efficient electron transport inhibitor myxothiazol **9** (Fig. 1) from cultures of the myxobacterial strain *Angiococcus disciformis* An d48 by mutating the loci encoding a negative regulatory protein via transposon mutagenesis (Sandmann et al. 2008). Myxothiacol **9** is an antimicrobial agent that inhibits the growth of Gram-positive eubacteria as well as the fungi *Mucor hiemalis, Trichophyton mentagrophytes* GBF 219, *Piricularia oryzae* GBF 158, and *Pythium debaryanum* GBF 161 with MIC values of 1–2, 0.02, 0.8, and 0.02 μg/ml, respectively (Gerth et al. 1980). Gratifyingly, Sandmann and coworkers were able to identify six different *A. disciformis* An d48 mutants with increased myxothiazol **9** production by as much as 8–33-fold compared to that of the wild-type strain. In addition, inactivating a negatively regulating gene, *thaA*, in *Burkholderia thailandensis* E264 isolated from rice paddies in Thailand produced higher yields of the highly unstable thailandamide A **10** and its stable geometric isomer thailandamide lactone **11** (Fig. 1). Compound **11** was evaluated in several bioassays and exhibited antiproliferative activity with a GI_{50} value of 28 μM as well as cytotoxic activity against human umbilical vein endothelial cells and HeLa cells (Ishida et al. 2010). Although the gene product of *thaA* is a quorum-sensing regulator homologous to LuxR regulators, it represents one of the few cases in which LuxR regulators downregulate genes involved in secondary metabolism.

Another direct way of activating the production of a metabolite is to directly manipulate transcription and translation using the "knock-in" promoter replacement strategy. Because secondary metabolite production in *Streptomyces* is primarily controlled at the level of transcription, the insertion of strong constitutive promoters in front of a cryptic gene cluster or regulatory gene has been successful in activating genes. In *Aspergillus oryzae*, the overexpression of a silent type III PKS gene cluster encoding *csyB* controlled by a strong α-amylase (*amyB*) constitutive promoter induced the production of the novel polyketide cspyrone B1 **12** (Fig. 1) (Seshime et al. 2010). In addition, Chiang and coworkers were able to induce the expression of the cryptic NR-PKS (AN1034.3) and HR-PKS (AN1036.3) gene clusters in *A. nidulans* once a citrinin-like biosynthesis transcriptional activator (*CtnR*) located in between the two PKSs was exchanged with the inducible alcohol dehydrogenase (*alcA*) promoter (Chiang et al. 2009). Gratifyingly, the new polyketide asperfuranone **13** (Fig. 1) was produced as a result of activating the expression of both HR-PKS and NR-PKS in *Aspergillus nidulans*. Asperfuranone **13** was later determined to exhibit antiproliferative activity against human non-small lung cancer A549 cells with an IC_{50} value of 15.3 μM (Wang et al. 2010). Furthermore, when the native promoters of 13 non-ribosomal peptide synthetase-like (NRPS-like) genes in *A. nidulans* were replaced by *alcA*, only the production of the known compound microperfuranone was produced and there were no changes in the remaining 12 genes, highlighting the limitations that can be encountered using this approach (Yeh et al. 2012). For other examples of how to engineer filamentous fungi to activate the production of cryptic metabolites in fungi, see the 2012 reviews by Lim et al. (2012) as well as Sanchez et al. (2012).

2.2 Heterologous Expression

Secondary metabolite production is contingent upon on numerous cellular processes that occur in a host with optimal genetics and biochemistry. By utilizing available genomic sequence data and molecular methods to reengineer hosts, researchers have been able to improve the overall yield of natural products via the heterologous expression of dedicated biosynthetic genes. The expression vector often requires a series of strong constitutive promoters to manipulate and integrate genes in the host's chromosomes for gene expression. Some gene clusters are quite large in size, requiring genes within the cluster to be stitched together. *Streptomyces coelicolor* has a large number of genetic tools available for manipulation, including several promoters for gene expression, as well as the ability to integrate large genomic libraries into its chromosomes (Gomez-Escribano and Bibb 2014). Importantly, *S. coelicolor* produces precursors for certain secondary metabolic pathways, as it already biosynthesizes several secondary metabolites. When choosing a host, the size of the genes and available promoters must also be considered for the optimal expression of a silent gene cluster as well as making sure there is a limited production of the host's own metabolites, which can interfere with the biosynthesis of the new natural product. For examples of *S. coelicolor* strains used in the heterologous expression of genes involved in natural product biosynthesis as well as their limitations, see the recent review by Gomez-Escribano and Bibb (2014).

In addition to using *S. coelicolor* as a host for heterologous gene expression, many studies have focused on engineering *E. coli*, which has a quick growth rate and extensive tools developed for its genetic manipulation. Importantly, *E. coli* has more understood primary metabolic pathways and lacks other endogenous secondary metabolic pathways that may interfere with the overexpressed biosynthetic pathway. Next, the choice of expression vector is often based on the quality of DNA, target genes, and screening strategy. For example, bromoalterochromide A **14** (Fig. 3), an antimicrobial and cytotoxic marine-derived non-ribosomal peptide (Speitling et al. 2007), was produced in *E. coli* by overexpressing 14 open reading frames of the 34-kb alterochromide gene locus (*alt*) as well as *altB-F* and *altG-M*, which overlap and are 6 and 25 kb, respectively (Ross et al. 2014). To the best of the authors' knowledge, this is the largest gene cluster encoding NRPS that has been successfully expressed as a single gene cluster in *E. coli*. This gene cluster encodes an NRPS, fatty acid synthase, flavin-dependent halogenase, and transporter proteins with all of the corresponding genes oriented in the same direction. *E. coli* was transformed with a pETDuet vector harboring 14 open reading frames and the *Pseudoalteromonas luteoviolacea* 2ta16 phosphopantetheinyl transferase (PPTase) gene in the second cloning site both under the control of the T7 promoter. Upon supplementing the fermentation media with potassium bromide, a 20-fold increase in the titer of bromoalterochromide A **14** was observed.

Additional engineered heterologous hosts are needed, as one major drawback to using *E. coli* is its lack of sufficient biosynthetic precursors; however, depending on the natural product, biosynthetic pathways can be introduced into *E. coli* or deleted to provide the required precursors. For example, overexpressing the large, multifunctional 300-kDa type I PKS [6-deoxyerythronolide B synthase (DEBS)] in reengineered *E. coli* (Pfeifer et al. 2001) produced 6-deoxyerythronolide B (6-DEB) **15**, the aglycone of the broad spectrum antibiotic erythromycin **16** (Fig. 3) (Wiley et al. 1957; McGuire et al. 1952). *E. coli* was genetically engineered in the following manner to produce 6-DEB **15**: (1) Three

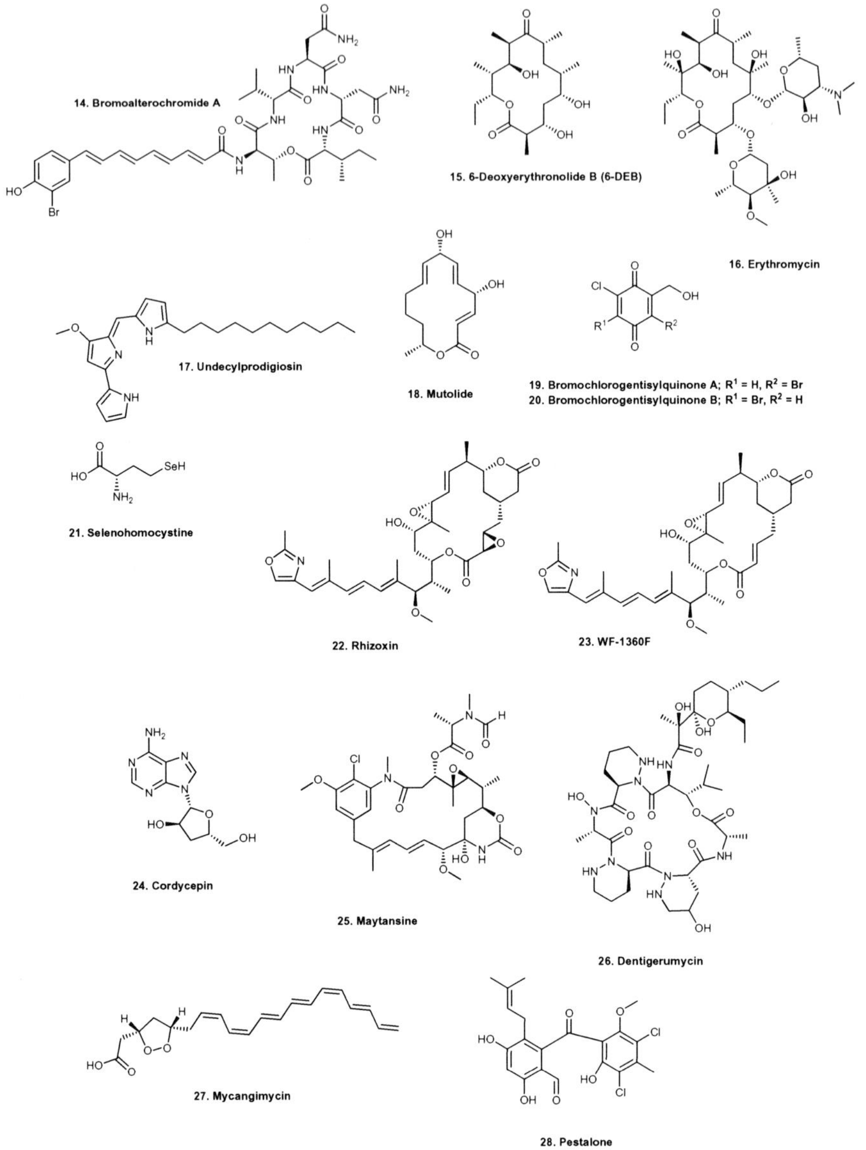

14. Bromoalterochromide A

15. 6-Deoxyerythronolide B (6-DEB)

16. Erythromycin

17. Undecylprodigiosin

18. Mutolide

19. Bromochlorogentisylquinone A; R^1 = H, R^2 = Br
20. Bromochlorogentisylquinone B; R^1 = Br, R^2 = H

21. Selenohomocystine

22. Rhizoxin

23. WF-1360F

24. Cordycepin

25. Maytansine

26. Dentigerumycin

27. Mycangimycin

28. Pestalone

Fig. 3   Structures **14–28**

DEBS genes were individually introduced from *S. erythraea* into *E. coli*; (2) the *Bacillus subtilis* phosphopantetheinyl transferase gene (*sfp*) under the control of a T7 RNA polymerase promoter was integrated via homologous recombination into the *E. coli* chromosome in the middle of the *prp* operon, inactivating the endogenous *prpRBCD* operon as well as ensuring phosphopantetheinylation of the seven acyl carrier protein domains in DEBS and accumulation of propionyl CoA and methylmalonyl CoA building blocks; and (3) overexpression of endogenous propionyl-CoA gene (*prpE* in BAP1) and (*birA*) genes also under the control of an IPTG-inducible T7 promoter. When genes were expressed at low temperatures, propionate was converted to 6-DEB **15**. A titer of 100 mg/l could be obtained once an accessory thioesterase TEII from *S. erythraea* was coexpressed, and *E. coli* were grown using a high-density fermentation procedure (Pfeifer et al. 2002). To read more about the additional modifications made to *E. coli* to increase the titer of 6-DEB **15**, see the review by Gao et al. (2010).

In order to assemble DNA into suitable gene expression vectors for the heterologous expression of biosynthetic gene clusters, the following modern techniques have been developed: sequence- and ligation-independent cloning (SLIC: Fig. 4) (Li and Elledge 2007), Gibson assembly ligation (Fig. 5) (Gibson et al. 2009), circular polymerase extension cloning (CPEC; Fig. 6) (Quan and Tian 2011), Golden Gate assembly method (Fig. 7) (Engler and Marillonnet 2011), and transformation-associated recombination (TAR; Fig. 8) (Larionov et al. 1996). These methods facilitate the assembly of large DNA fragments with fewer cloning steps and circumvent the sequence requirements of traditional cloning methods. In some cases, there is no need to initially construct and screen genomic libraries of random DNA fragments and genes can be integrated into a host genome for stable expression, which is useful if there is no plasmid available for the host of choice. See Figs. 4, 5, 6, 7, and 8 for overviews of these cloning strategies.

SLIC (Fig. 4) is a method that relies on in vitro homologous recombination to construct multifragment plasmids in which an insert is prepared with a primer with a 20-bp or longer extension homologous to each end of the linearized destination vector (Li and Elledge 2012). In a one-step, room temperature reaction, both insert and linearized vector are incubated with T4 DNA polymerase in the absence of deoxyribonucleotide triphosphates (dNTPs) to exploit its 5′ exonuclease activity to generate single-stranded DNA overhangs. Exonuclease activity is terminated by the addition of deoxycytidine triphosphate (dCTP), switching the function of T4 DNA polymerase to that of a polymerase. T4 DNA polymerase eventually stalls without additional dNTPs, resulting in the homologous ends of the insert and vector annealing in vitro. Upon transformation into *E. coli*, natural repair processes remove any single-stranded DNA gaps, resulting in circular plasmids that can harbor up to six fragments of DNA.

For constructing plasmids with more than six DNA fragments (i.e., several hundreds of kilobases), the Gibson assembly ligation and CPEC methods are often used. The Gibson assembly ligation method (Fig. 5) is an in vitro, isothermal, one-step reaction that rapidly assembles DNA (i.e., inserts and linearized destination vector) with 40 bp of overlapping DNA. A 5′-exonuclease (e.g., 5′ T5 exonuclease) recesses the double-stranded DNA of the inserts and vector from the 5′ ends until its inactivation over the course of the incubation. Then, the complementary single-stranded 3′ overhangs subsequently anneal, DNA polymerase (e.g., Phusion polymerase) fills the gaps between anneal fragments with dNTPs, and DNA ligase (e.g., *Taq* ligase) repairs the single-stranded nicks in DNA.

SLIC

1. PCR with ≥ 20-bp primer with homologous sequence to produce the following:

2. Add T4 DNA polymerase (with 5' exonuclease activity), no dNTPs, linearized vector, and insert(s)

3. Add dCTPs and annealing occurs

4. Transformation into *E. coli* removes gaps

Fig. 4 SLIC methodology

Gibson assembly ligation

1. PCR with ≥ 20-bp primer with homologous sequence yields inserts with appropriate ends for recombination:

2. Add T5 exonuclease, DNA polymerase, dNTPs, and DNA ligase to linearized vector and insert(s)

Complementary single-stranded overhang anneals

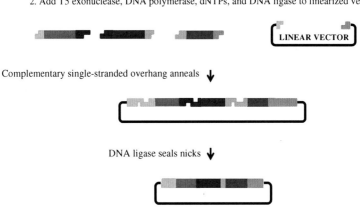

DNA ligase seals nicks

Resulting plasmid is ready for transformation into host

Fig. 5 Gibson assembly ligation methodology

Circular polymerase extension cloning

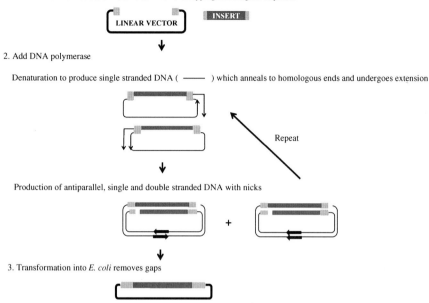

1. Double stranded insert and linear vector with overlapping homologous sequence

2. Add DNA polymerase

Denaturation to produce single stranded DNA (————) which anneals to homologous ends and undergoes extension

Repeat

Production of antiparallel, single and double stranded DNA with nicks

3. Transformation into *E. coli* removes gaps

Fig. 6 Circular polymerase extension cloning (*CEPC*)

This reaction can take place in under 1 h and assemble PCR products as large as 583 kb or ligate up to 300-kb of PCR products in *E. coli* expression vectors (Gibson et al. 2009). An added benefit to using this method is that ligated DNA assembly products are more efficiently electroporated into *E. coli* compared to their unligated counterparts.

For high-throughput, combinatorial, or multifragment cloning, the CPEC (Fig. 6) method is a one-step, in vitro reaction used to assemble and clone DNA using a single polymerase (Quan and Tian 2014). The polymerase extension method is used to join overlapping regions of the insert and linear vector. The double-stranded insert and linearized vector are denatured, and the resulting single strands of DNA anneal with their overlapping ends and extend using each other as a template to form double-stranded circular plasmids with only two nicks, one on each single strand. No restriction digestion, ligation, or single-stranded homologous recombination is required, which can be extremely useful when cloning complex genome libraries. Lastly, the nicks are covalently closed upon transformation into *E. coli* using its natural repair processes.

The Golden Gate assembly (Fig. 7) method is another quick one-pot method involving the restriction ligation of an insert and vector to form seamless recombinant plasmids without unwanted DNA sequences that lead to additional amino acids attached to the protein of interest (Engler and Marillonnet 2011). In this method, type IIS restriction endonucleases (e.g., *Bsa*I) cleave DNA distal to its recognition sequence flanking both the insert and ends of the linearized destination vector,

Golden Gate assembly method

1. Use PCR to add two BsaI recognition sites pointing inwards in insert and use linearized vector with 2 BsaI sites

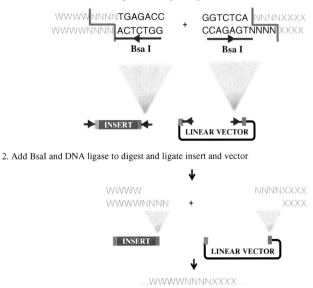

2. Add BsaI and DNA ligase to digest and ligate insert and vector

Resulting plasmid is ready for transformation into host

Fig. 7   Golden gate assembly methodology

allowing 4-bp overhangs to be customized for a seamless assembly, while DNA ligase connects the DNA fragments. Two DNA ends are flanked by type IIS restriction sites such that digestion produces ends with complementary 4-bp overhang, which can be ligated, creating a junction that lacks the original restriction site. Importantly, this method does not depend on the specific sequence of each gene of interest, and it can be used to assemble or shuffle several DNA fragments as long as they are flanked by the correct 4-bp overhang sequence highlighted in blue or pink in Fig. 7.

One of the most popular one-step cloning approaches to ligate large, functionally intact genes involved in natural product biosynthesis is TAR (Fig. 8), which relies on native homologous recombination in *Saccharomyces cerevisiae* to insert DNA of a specific sequence into a "capture" vector harboring a yeast artificial chromosome replication sequence (e.g., *CEN6*), selection marker (e.g., *HIS3*), and homologous arms (approximately 60 bp of homology to the insert) (Kouprina and Larionov 2008). The homologous arms are essentially gene-targeting sequences. Upon transformation in *S. cerevisiae* spheroblasts, large natural product gene clusters or genomic DNA fragments with overlapping homologous sequences to the linearized capture vector are covalently linked to the vector from a pool of DNA without having to screen or construct genomic DNA libraries. When cloning DNA

Transformation-associated recombination

1. Capture vector has gene sequence of 40–60-bp homologous sequence to target inserts:

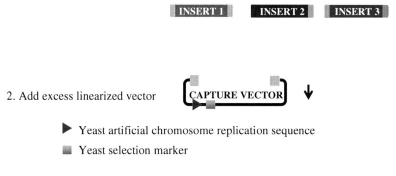

2. Add excess linearized vector

▶ Yeast artificial chromosome replication sequence

▪ Yeast selection marker

3. Homologous recombination occurs upon transformation into *S. cerevisiae*

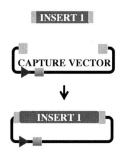

4. Positive colonies with gene of interest can be selected using the selection marker (▪)

Fig. 8 Transformation-associated recombination (*TAR*) methodology

from metagenomic samples, cosmid or fosmid libraries containing 30–40 kb are first generated and then large gene clusters are subsequently cloned using TAR. However, when using this method, the cloned gene cluster cannot be subsequently engineered. For examples of how TAR has been used to assemble natural product gene clusters, see recent papers from the Brady research group (Kim et al. 2010; Feng et al. 2010; Yamanaka et al. 2014).

Although the methods described are powerful, they require careful planning and typically are less error prone with automation algorithms and software packages [e.g., Clotho framework (Densmore 2009) and Eugene language (Bilitchenko et al. 2011)]. Using these methods, the focus of heterologous gene expression has now shifted from increasing the production of a novel metabolite to finding new secondary metabolites on an industrial scale. Other methods used for heterologous gene expression are BioBrinks^{TM} assembly and variants (Shetty et al. 2008),

Red/ET recombination (Rivero-Müller et al. 2007), *ori*T-directed capture, seamless ligations cloning extract (SLICE) (Zhang et al. 2012), and DNA assembly in *Bacillus* using the domino method (Itaya et al. 2008). For more details on the methods used for heterologous gene expression, see the reviews by Ellis et al. (2011), Zotchev et al. (2012), as well as Ongley et al. (2013).

2.3 Gene Activation by the Addition of Rare Earth Metals and Other Additives

An inexpensive method of activating cryptic gene clusters that does not require genetic engineering or sequence information is the supplementation of fermentation media with various additives. Changes in microbial metabolomes can be observed as microorganisms respond to stress stimuli, such as changes in pH, temperature, and the addition of metals, antibiotics, and flavonoids. While studying the effects of rare earth metals on bacterial physiology, Kawai and coworkers observed an increase in the expression of genes involved in antibiotic production when the fermentation media of *S. lividans* were supplemented with trace amounts of scandium (Kawai et al. 2007). Antibiotic production increased by twofold to 25-fold, and the expression of silent genes involved in the production of actinorhodin **4** (Fig. 1) was activated due to the effects of scandium at the level of transcription of biosynthetic pathway-specific, positive regulatory genes (e.g., *actII-ORF4*). Years later, Tanaka et al. added trace amounts of rare earth metals (e.g., scandium and/or lanthanum) in the fermentation broth of *S. sioyaensis*, activating the expression of nine genes that were previously silent or poorly expressed in *S. coelicolor* A3(2) (Tanaka et al. 2010). Notably, gene expression increased by 2.5- to 12-fold. Although, the mechanism of action of scandium and other rare earth metals remain unknown, the authors speculate that microorganisms may have developed the ability to respond to low levels of these elements as they are deposited throughout the world over time. For more information on the biological effects of rare earth metals in microorganisms, see the recent review by Ochi et al. (2014).

Enzyme inhibitors have also been added to fermentation media to trigger the expression of cryptic biosynthetic pathways. For example, histone deacetylase (HDAC) inhibitors derepress biosynthetic pathways in fungi and modify the expression of genes involved in secondary metabolism in a variety of species of *Streptomyces* (Moore et al. 2012). Several HDAC inhibitors, such as sodium butyrate, MS-275, and valproic acid, have been reported to induce the expression of gene clusters involved in the production of bioactive, pigmented secondary metabolites, such as actinorhodin **4** (Fig. 1) and undecylprodigiosin **17** (Fig. 3), in *S. coelicolor* grown on minimal medium. Actinorhodin **4** exhibits antimicrobial activity (Wright and Hopwood 1976), and undecylprodigiosin **17** exhibits antimicrobial, antioxidative, and cytotoxic activities (Ho et al. 2007; Stankovic et al. 2012). Under standard fermentation conditions, most of the biosynthetic pathways in *S. coelicolor* are silent. However, upon the addition of HDAC inhibitors, concomitant changes were observed in chromatin structure with an upregulation of specific biosynthetic pathways. Genes involved in transcriptional control are speculated to be more accessible in the nucleoid, and upon the deacetylation of histone proteins, the chromatin changes shape, inducing the expression of biosynthetic pathways via mechanisms that are currently being investigated.

Other enzyme inhibitors have been reported to increase the production of secondary metabolites. For example, out of the 30,569 small molecules that comprise the Canadian Compound Collection, 19 small molecule inhibitors of fatty acid biosynthesis in *S. coelicolor* increased the production of secondary metabolites, specifically pigments (Craney et al. 2012). In addition, using the one strain–many compounds (OSMAC) approach, in which the cultivation parameters (e.g., media composition, addition of enzyme inhibitors, and aeration) of a single microbial strain are systematically altered, Bode and coworkers isolated the rare compound mutolide **18** (Fig. 3) from *Sphaeropsidales* sp. F-24′707y upon the addition of tricyclozole, an inhibitor of 1,8-dihydroxynapthalene biosynthesis (Bode et al. 2002). The weakly antibacterial mutalide **18** was originally produced from a *Sphaeropsidales* sp. F-24′707y mutant, demonstrating the potential for identifying new naturally occurring molecules by blocking distinct enzymatic steps in a biosynthetic pathway (Pettit 2011). Furthermore, inhibitors of competing biosynthetic pathways, such as small molecules that inhibit FabI, a gene involved in fatty acid biosynthesis, have also been reported to improve the production of secondary metabolites (Craney et al. 2012).

The addition of high concentrations of *N*-acetylglucosamine, a major component of the eubacterial cell wall and chitin monomer, to fermentation media has been shown to induce gene expression by mimicking the autolytic degradation of the vegetative mycelium, a major checkpoint in activating secondary metabolism (Rigali et al. 2008). *N*-acetylglucosamine induces the expression of antibiotic pathway-specific activators via the transcriptional repressor *DasR*, stimulating antibiotic production and sporulation under nutrient-depleted conditions. In the presence of this additive, Rigali and coworkers observed the production of the antibiotic undecylprodigiosin **17** (Fig. 3) and actinorhodin **4** (Fig. 1) by *S. coelicolor* grown in minimal media. Other species of *Streptomyces*, such as *S. clavuligerus*, *S. collinus*, *S. griseus*, *S. hygroscopicus*, and *S. venezuelae*, have also been observed to increase antibiotic production when grown in minimal media, suggesting that *N*-acetylglucosamine may stimulate antibiotic production in a variety of streptomycetes.

Supplementing the fermentation medium of marine extremophiles with inorganic halides, such as $CaBr_2$ or inorganic fluoride, has also activated the production of new metabolites. Nenkep and coworkers observed that the fungus *Phoma herbarum*, isolated from the marine red alga *Gloiopeitis tenax* in Cheeokpo, Tongnyeong, Korea, did not produce halogenated metabolites under saline conditions (Nenkep et al. 2010). Therefore, they supplemented the fermentation medium with $CaBr_2$ and isolated the halogenated benzoquinones bromochlorogentisylquinones A and B (**19–20**) (Fig. 3). Compounds **19–20** are stereoisomers that exhibited 2,2-diphenyl-1-picrylhydrazyl (DPPH) radical scavenging activity with IC_{50} values of 3.8 and 3.9 μM, respectively, which were lower concentrations compared to that of the L-ascorbic acid-positive control (IC_{50}, 20 µM).

Precursor-directed biosynthesis is another strategy that uses already existing genetic machinery to produce new novel compounds. For example, increasing the amount of acetyl-CoA, which is used in both primary and secondary metabolism, can improve the yields of low-yielding secondary metabolites. Other precursors have been shown to produce new natural products in mutant strains. Imada et al. supplemented the minimal medium used to ferment a selenomethionine-resistant strain of the Gram-positive marine-derived *Bacillus* sp. No 14 with seleno-DL-methionine or seleno-DL-ethionine and observed the production of a previously unknown antibiotic (Imada et al. 2002). The active component was determined to be selenohomocystine **21** (Fig. 3), which was produced by the conversion of the eubacterial products selenomethionine and selenoethionine. Compound **21** exhibited antibacterial activity against *Micrococcus luteus* IFO 3333 with an MIC of 1.5 µg/ml. For other examples of altered fermentation conditions producing cryptic microbial metabolites, see the review by Bode et al. (2002).

2.4 Gene Activation via the Cocultivation of Microorganisms

Although the standard method for cultivating microorganisms is to ferment a single microbial strain, the growth conditions of this strain in its natural environment in which it interacts with other microorganisms are remarkably different. Molecules analogous to eubacterial quorum-sensing and other elicitors are thought to be involved in activating the expression of cryptic biosynthetic gene clusters. Notably, eubacterial–fungal (Schroeckh et al. 2009; Oh et al. 2007) and fungal–fungal (Zhu and Lin 2006) interactions have been reported to activate genes involved in the production of secondary metabolites, such as polyketides and non-ribosomal peptides, required for protection (Giddings and Newman 2013). Thus, by mimicking the natural ecological conditions in which microorganisms dwell in complex microbial communities and creating a competitive environment in which microbes compete for limited resources, various microbial defense mechanisms can be activated that involve the production of secondary metabolites. Figure 9 shows some of the physical differences that can be observed in a liquid coculture (Fig. 9a) compared to monocultures as well as the inhibition zones that result from the competition between various microorganisms (Fig. 9b–c).

Over the last two decades, there has been a significant increase in the number of reports of microorganisms living in mutualistic environments that produce valuable therapeutic agents. For the purpose of this book, microorganisms that inhabit mutualistic environments, such as endosymbionts in plants, insects, and other animals that are either unculturable or unable to produce specific natural products outside of these environments, are considered to be extremophiles. The small molecules produced by these extremophiles can serve as eubacterial quorum-sensing agonists to produce new bioactive molecules. For example, the production

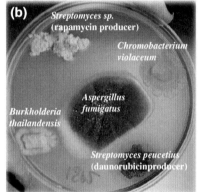
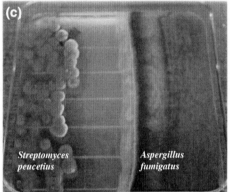

Fig. 9 Liquid coculture experiments

of rhizoxin **22** (Fig. 3) is a prime example of successful cocultivation to produce new natural products. *Burkholderia rhizoxinica*, the intracellular symbiont of the pathogenic fungus *Rhizopus* sp., is required for fungal production of the antifungal and cytotoxic agent rhixoxin **22** via the production of its precursor WF-1360F **23** (Partida-Martinez and Hertweck 2005). Other mutualistic symbionts have been required for the production of the following bioactive natural products in a host: cordycepin **24** (Masuda et al. 2006), maytansine **25** (Cassady et al. 2004; Wings et al. 2013), dentigerumycin **26** (Oh et al. 2009), and mycangimycin **27** (Scott et al. 2008) (Fig. 3).

Not only can host–symbiont interactions induce the production of new metabolites, but interspecies cross talk between symbionts or simply two or more microorganisms can also increase the production of cryptic metabolites, known compounds, and related analogs. Recently, several marine-derived extremophiles have been used in mixed fermentation studies to produce novel metabolites, as marine invertebrates are known to consist of symbiotic microorganisms that produce a wide array of chemically diverse and bioactive molecules. Cueto and coworkers isolated a new chlorinated antibiotic produced via the cocultivation of the marine-derived fungus *Pestalotia* sp. strain CNL-365 and an antibiotic-resistant, marine-derived *Thalassopia* sp. (CNJ-328) (Cueto et al. 2001). The compound was determined to be the benzophenone pestalone **28** (Fig. 3), which exhibited potent antibacterial activity against methicillin-resistant *Staphylococcus aureus* and vancomycin-resistant *Enterococcus faecium* with MIC values of 37 and 78 ng/ml, respectively. This compound also exhibited cytotoxic activity against the National Cancer Institute (NCI)-60 cell line screen with an average GI_{50} value of 6 μM. The authors speculate that pestalone **28** is a fungal product, as its methylated analog has been isolated from another fungus. Based on the antibiotic activity of pestalone **28**, this compound should be evaluated in other infectious disease models.

The Fenical research group at Scripps Institution of Oceanography reported the induction of four new pimarane diterpenoids via the cocultivation of marine-derived α-proteobacterium *Thalassopia* sp. (CNJ-328) and fungus *Libertella* sp. CNL-523 strain (Oh et al. 2005). Libertellenones A–D (**29–32**) (Fig. 10) were isolated and determined to be pimarane diterpenoids with a variation of hydroxylation and acetylation. Libertellenones A–D (**29–32**) exhibited cytotoxicity against human colorectal carcinoma HCT 116 cells with IC_{50} values of 15, 15, 53, and 0.76 μM, respectively. The marine-derived fungus was isolated from an unidentified ascidian collected in the Bahamas, and the α-proteobacterium was isolated as a contaminant of the fungal culture, which also produces pestalone **28**. Interestingly, this is the second example of pestalone **28**, most likely produced from marine-derived fungi, being induced via mixed fermentation of a eubacterial strain.

From cocultures of two unidentified mangrove-derived endophytic fungi isolated from the South China Sea, Zhu and coworkers identified marinamide, a novel cytotoxic pyrrolyl 1-isoquinolone alkaloid, and its methyl ester derivative (Zhu and Lin 2006). In 2013, the authors later revised the structures of the marinamides (**33–34**) (Fig. 10) to pyrrolyl 4-quinolones, which represent a new chemical class (Zhu et al. 2013). Marinamide **33** is a pyrrolyl 4-quinoline that exhibited cytotoxic activity against the human liver hepatocellular carcinoma HepG2, lung cancer 95-D, gastric cancer MGC832, and HeLa cells with IC_{50} values of 7 nM, 0.4 nM, 91 nM, and 529 nM, respectively. The marinamide methyl ester **34** exhibited cytotoxicity against human liver hepatocellular carcinoma HepG2, lung cancer 95-D, gastric cancer MGC832, and HeLa cells with IC_{50} values of 2.52 μM, 1.54 μM, 13 nM, and 110 nM, respectively. These results highlight that slight modifications to natural products can significantly alter their biological activity. Furthermore, the data suggest that pyrrolyl 4-quinolones should be further developed to find new cancer therapeutics.

The Fenical research group also isolated novel cyclic depsispeptides, emericellamides A and B (**35–36**) (Fig. 10), from the mixed fermentation of marine-derived fungus *Emericilla* sp. strain CNL-878 and *Salinispora arenicola* strain CNH-665 (Oh et al. 2007). *Emericella* sp. (CNL-878) was isolated from the surface of green alga of the *Halimeda* genus collected at Madang Bay in Papua New Guinea, and *S. arenicola* (CNH-665) was isolated from a sediment obtained from the Bahamas. Emericellamides A and B (**35–36**) are cyclic peptides containing 3-hydroxy-2,4-dimethyldecanoic and 3-hydroxy-2,4,6-trimethyldodecanoic

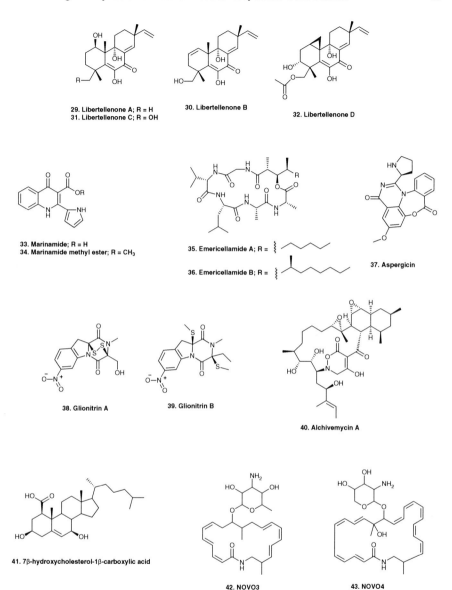

Fig. 10 Structures **29–44**

acids, respectively, and both metabolites are of mixed non-ribosomal peptide and polyketide origin. Emericellamide A **35** exhibited antimicrobial activity against methicillin-resistant *S. aureus* with an MIC value of 2.3 µg/ml but weak cytotoxicity against human colorectal carcinoma HCT 116 cell line with an IC$_{50}$ value of 23 µM. On the other hand, emericellamide B **36** exhibited weaker antimicrobial activity against methicillin-resistant *S. aureus* with an MIC value of 3.9 µg/ml and even weaker cytotoxicity with an IC$_{50}$ value of 40 μM against the human colorectal carcinoma HCT 116 cell line.

Based on marine-derived mangrove fungi being an abundant source of new chemical entities, Zhu and coworkers searched for new metabolites from mixed fermentation involving marine-derived fungi (Zhu et al. 2011). Gratifyingly, the authors identified a new antibacterial alkaloid produced via the cofermentation of two strains of marine-derived mangrove epiphetic fungi *Aspergillus* sp. isolated from rotten fruit from the mangrove *Avicennia marina* in the South China Sea. The antibiotic was determined to be the benzodiazepine aspergicin **37** (Fig. 11), which was not

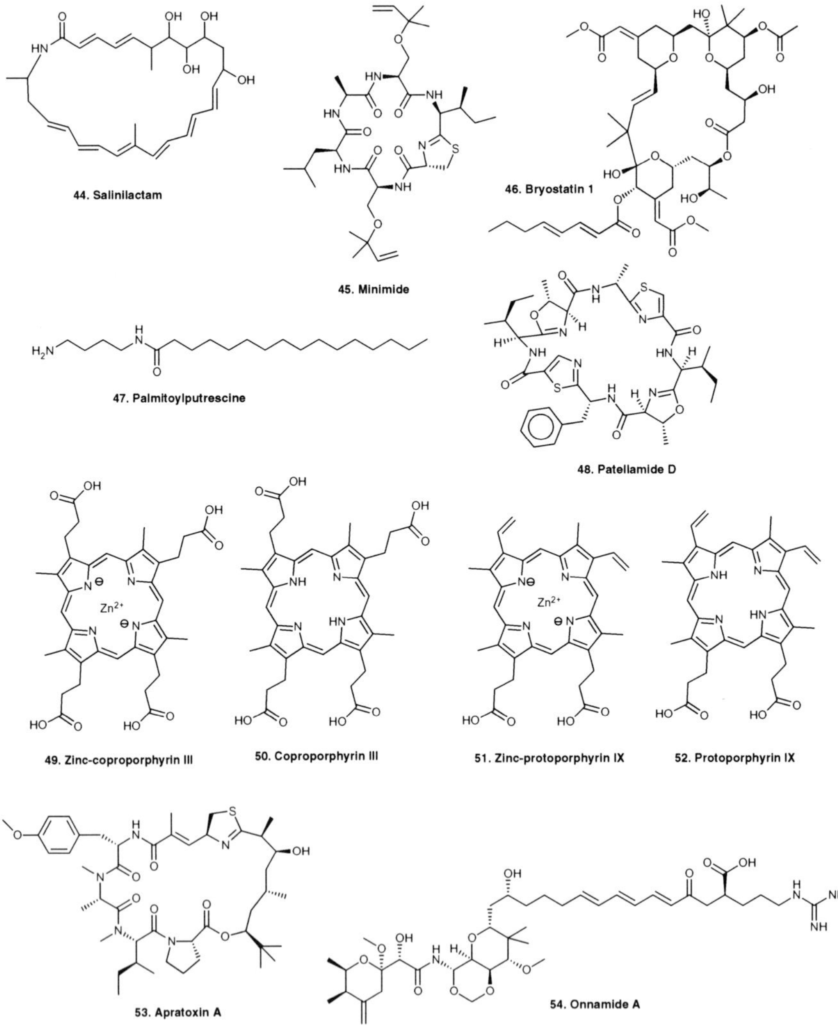

Fig. 11 Structures **44–54**

produced in monocultures of either fungus. This compound exhibited mostly weak antimicrobial activity against *S. aureus* (MIC, 62.5 µg/ml), *S. epidermidis* (MIC, 31.25 µg/ml), *B. subtilis* (MIC, 15.62 µg/ml), *B. dysenteriae* (MIC, 15.62 µg/ml), *B. proteus* (MIC, 62.50 µg/ml), and *E. coli* (MIC, 31.25 µg/ml).

An interesting eubacterium from mine drainage, *Sphingomonas* sp. KMK-001, was cocultured with another mine-drainage-derived fungus, *Aspergillus fumigatus* KMC-901, and reported to produce a new metabolite, glionitrin A **38** (Fig. 10) (Park et al. 2009). These microorganisms were isolated from an acidic mine drainage enriched with heavy metals collected from a horizontal pit 750 m high from the abandoned Young-Dong coal mine in Ganneung, South Korea. Glionitrin A **38**, a diketopiperazine disulfide that contains a nitro aromatic ring, exhibited potent antimicrobial activity against strains of methicillin-resistant *S. aureus* and *M. luteus* IFC12708 with MIC values of 0.78 µg/ml. In addition, compound **38** exhibited antimicrobial activity against *B. subtilis* (ATCC6633; MIC, 6.25 µg/ml), *Proteus vulgaris* (ATCC3851; MIC, 3.13 µg/ml), *Salmonella typhimurium* (ATCC14028; MIC, 3.13 µg/ml), *A. fumigatus* (HIC6094; MIC, 12.5 µg/ml), and *Trichophyton rubrum* (IFO9185; MIC, 12.5 µg/ml). Also, glionitrin A **38** exhibited potential to moderate cytotoxic activity against the human colorectal carcinoma HCT 116, lung carcinoma A549, gastric adenocarcinoma AGS, prostate cancer DU145, breast adenocarcinoma MCF7, and hepatocellular carcinoma HepG2 cancer cell lines with IC_{50} values of 0.82, 0.55, 0.45, 0.25, 2.0, and 2.3 µM, respectively. Two years later, glionitrin B **39** (Fig. 10), a glionitrin A **38** analog lacking the disulfide bond, was reported to be produced from the same coculture (Park et al. 2011). Glionitrin B **39** was not cytotoxic against the prostate cancer DU145 cell line; however, 60 µM of the compound reduced the invasiveness of prostate cancer DU145 cells by 46 %, which is an important step in tumor metastasis. Understanding the anti-invasive mechanisms of glionitrin B **39** may aid in the development of a new antimetastatic agent for the treatment of cancer.

Coculturing various species of *Streptomyces* with mycolic acid-producing eubacteria has also been reported to induce the biosynthesis of cryptic gene clusters by Onaka et al. (2011). The production of red pigments, actinorhodin **4** (Fig. 1) and undecylprodigiosin **17** (Fig. 3), by *S. lividans* TK23 was induced via direct interaction with the mycolic acid-containing *Tsukamurella pulmonis* TP-B0596. Cocultures with *T. pulmonis* modified the production of other natural products in 88.4 % of soil-derived *Streptomyces* strains. Other mycolic acid-containing eubacteria, *Rhodococcus erythropolis* and *Cornyebacterium glutamicum*, were also observed to increase and decrease natural product biosynthesis in 87.5 and 90.2 % strains of *Streptomyces* studied, respectively. Notably, *S. endus* S-522 produced the new antibiotic alchivemycin A **40** (Fig. 10) when cocultivated with *T. pulmonis*. This compound exhibited antibiotic activity against *M. luteus* at a concentration of 0.06 µg/ml. In addition, alchivemycin A **40** inhibited the invasion of murine colon carcinoma 26-L6 cells into Matrigel with an IC_{50} value of 0.34 µM (Igarashi et al. 2010). The authors speculate that the localization of mycolic acid to the outer cell layer of the inducer eubacterium influences secondary metabolism in *Streptomyces* through direct interactions.

Table 1 Bioactive compounds discussed and their producing microorganisms

Microorganism	Compound	Bioactivity	References
Salinispora tropica	Fluorosalinosporamide A 1	20S Proteasome inhibitor	Eustáquio and Moore (2008)
S. tropica	Salinosporamide X7 2	20S Proteasome inhibitor	Eustáquio, and. Moore (2008), Nett et al. (2009)
S. tropica	Salinosporamide A 3	20S Proteasome inhibitor	Feling et al. (2003)
Streptomyces coelicolor A3(2)	Actinorhodin 4	Antimicrobial	Wright and Hopwood (1976)
S. ambofaciens	Stambomycins A–D (5-8)	Antimicrobial; cytotoxic	Laureti et al. (2011)
Angiococcus disciformis An d48; Myxococcus fulvus	Myxothiazol 9	Electron transport inhibitor; antimicrobial	Sandmann et al. (2008), Gerth et al. (1980)
Burkholderia thailandensis E264	Thailandamide A 10	N/A	Ishida et al. (2010)
B. thailandensis E264	Thailandamide lactone 11	Cytotoxic; antiproliferative	Ishida et al. (2010)
Aspergillus oryzae	Cspyrone B1 12	N/A	Seshime et al. (2010)
A. nidulans	Asperfuranone 13	N/A	Chiang et al. (2009)
Pseudoalteromonas maricaloris strain KMM 636T; heterologous expression in Escherichia coli	Bromoalterochromide A 14	Antimicrobial; cytotoxic	Speitling et al. (2007)
Saccharopolyspora erythraea; heterologous expression in E. coli; S. erythreus	6-Deoxyerythronolide B 15	N/A	Pfeifer et al. (2002), McGuire et al. (1952), Flynn et al. (1954), Martin and Rosenbrook (1967)
S. erythraea; S. erythreus	Erythromycin 16	Antimicrobial	McGuire et al. (1952), Flynn et al. (1954), Wiley et al. (1957)
Streptomyces sp.; Serratia sp.	Undecylprodigiosin 17	Cytotoxic; antimicrobial; antioxidant	Ho et al. (2007), Stankovic et al. (2012)
Sphaeropsidales sp. F-24'707y and mutant	Mutolide 18	Weak antimicrobial	Bode et al. (2002), Pettit (2011)
Phoma herbarum fermentation medium supplemented with CaBr2	Bromochlorogentisylquinones A and B (19–20)	Antioxidant	Nenkep et al. (2010)

(continued)

Table 1 (continued)

Microorganism	Compound	Bioactivity	References
Bacillus sp. no 14	Selenohomocystine **21**	Antimicrobial	Imada et al. (2002)
Burkholderia sp. and *Rhizopus* sp. in *Oryza sativa* host	Rhizoxin **22**	Cytotoxic; phytotoxin	Iwasaki et al. (1984), Partida-Martinez and Hertweck (2005, 2007)
Cordyceps militaris; mutant caterpillar fungus strain NBRC 9787 in Lepidopteron larvae and the pupae of insects or spiders	Cordycepin **23**	Antimalarial; cytotoxic; antimicrobial, anti-inflammatory, antioxidant; antidepressant	Cunningham et al. (1950), Bentley et al. (1951), Sakurai and Masuda (2013), Tuli et al. (2014)
Burkholderia sp. in *Rhizopus* sp.	WF-1360F **24**	Cytotoxic; phytotoxin	Iwasaki et al. (1984), Partida-Martinez and Hertweck (2005, 2007)
Actinosynnema pretiosum ssp. *Auranticum* in *Maytenus ovatus*; *M. buchananii*; *Putterlickia verrucosa*	Maytansine **25**	Antiparasitic; antimicrobial	Kupchan et al. (1972), Wings et al. (2013)
Pseudonocardia sp. in *Apterostigma dentigerum*	Dentigerumycin **26**	Antimicrobial	Oh et al. (2009)
Entomocorticium sp. A in the presence of *Ophiostoma minus* in *Dendroctonus frontalis*	Mycangimycin **27**	Antimalarial; antimicrobial	Scott et al. (2008), Oh et al. (2009)
Cocultures of *Pestalotia* sp. strain CNL-365 and an unidentified antibiotic-resistant marine bacterium (CNJ-328)	Pestalone **28**	Antimicrobial	Cueto et al. (2001)
Cocultures of *Acremonium* sp. CNJ-328 and *Libertella* sp. CNL-523 strain	Libertellenones A–D (**29–32**)	Cytotoxic	Oh et al. (2005)
Cocultures of two unidentified endophytic fungi isolated from the marine-derived mangrove	Marinamides A–B (**33–34**)	Cytotoxic	Zhu and Lin (2006)

(continued)

Table 1 (continued)

Microorganism	Compound	Bioactivity	References
Cocultures of *Emericilla* sp. strain CNL-878 and *S. arenicola* strain CNH-665	Emericellamides A and B (**35–36**)	Antimicrobial; weakly cytotoxic	Oh et al. (2007)
Cocultures of marine-derived mangrove epiphytic *Aspergillus* sp. isolated from the mangrove fruit *Avicennia marina*	Aspergicin **37**	Antimicrobial	Zhu et al. (2011)
Cocultures of *Sphingomonas* sp. KMK-001 and *A. fumigatus* KMC-901	Glionitrins A–B (**38–39**)	Antimicrobial (**37**); cytotoxic (**37**); inhibitor of the invasiveness of prostate cancer DU145 (**38**)	Park et al. (2009, 2011)
Cocultures of *S. endus* S-522 *fumigatus* KMC-901 and *Tsukamurella pulmonis* TP-B0596	Alchivemycin A **40**	Antimicrobial; inhibitor of the invasiveness of murine colon carcinoma 26-L6 cells into Matrigel	Onaka et al. (2011), Igarashi et al. (2010)
Cocultures of marine alga-derived *Aspergillus* sp. BM05 and an unidentified bacterium (BM05BL)	7*β*-Hydroxycholesterol-1*β*-carboxylic acid **41**	Cytotoxic	Ebada et al. (2014)
P1532 soil-derived strain of *Streptosporangium* (ATCC PTA-8676)	NOVO3 **42**	Antimicrobial	Lewis et al. (2012)
Z0363 (NRRL 50107) strain of *Amycolatopsis*	NOVO4 **43**	Antimicrobial	Lewis et al. (2009)

N/A information not available

More reports will be published on the cofermentation of microorganisms, increasing the number of compounds listed in Table 1, as roughly or so dozen papers have already been published between 2013 and 2014. In early 2014, an entire issue of Marine Drugs (Volume 12, Issue 2) was dedicated to this topic. Moreover, in March of 2014, Ebada and coworkers reported the use of cofermentation of two marine alga-derived microorganisms to produce a new steroid 7β-hydroxycholesterol-1β-carboxylic acid **41** (Fig. 10). Along with the known 7β-hydroxycholesterol, 7α-hydroxycholesterol, and ergosterol-5α,8α-peroxide, compound **41** exhibited moderate-to-weak cytotoxic activity against human myelogenous leukemia K562, colon colorectal carcinoma HCT 116, cisplatin-resistant ovarian carcinoma A2780CisR, and ovarian carcinoma A2780 cells with IC$_{50}$ values ranging from 10 to 100 μM (Ebada et al. 2014). For more examples of the use of cocultivation in natural product discovery, see the excellent review by Marmann et al. (2014) as well as the chapter on fungal actinomycete interactions in the recent book, "Natural Products, Discourse, Diversity and Design" edited by Osbourn, Goss, and Carter (Schroeckh et al. 2014).

In addition to cofermentation, Kaeberlein and coworkers have shown that two or more uncultivatable organisms can also be grown separately in a simulated native environment, such as a semiporous culture chamber, allowing nutrients and growth factors, not cells, to cross a porous membrane to activate growth and the production of secondary metabolites (Kaeberlein et al. 2002). The same research group cultured up to 40 % of species from marine biofilms as well as aerobic and anerobic species from subsurface sediment (Bollmann et al. 2010) in this manner and commercially patented the new secondary metabolites produced, such as the glycosylated macrolactam antibiotics NOVO3 **42** and NOVO4 **43** (Fig. 10), which exhibit potent antibacterial activity against methicillin-resistant *S. aureus* and vancomycin-resistant *Enterococci* with minimal in vitro toxicity to mammalian cells (Lewis et al. 2009, 2012). Microbial isolates have also been grown in simulated environments, such as membranes suspended on soil slurries, and single cells have been embedded in agarose microdroplets and cultured directly in seawater (Zengler et al. 2002) or on living coral (Ben-Dov et al. 2009). Thus, remarkable developments are being made using culture-based and culture-independent methods, and a combination of these methods will be useful in identifying novel pharmacophores from microbes inhabiting different ecosystems in the future.

3 Bioprospecting Metagenomes for Bioactive Agents

Whole-genome sequencing in the mid-1990s paved the way for the field of genomics, which has enabled scientists to link phenotypes to underlying genetic mechanisms. Within the last 10 years, new sequencing technologies have emerged, facilitating the rapid production of microbial genomic data and full characterization of gene products in unexplored biosynthetic pathways. Rapidly evolving next-generation sequence (NGS, post-Sanger sequencing) technologies have led to the quick and inexpensive high-throughput sequencing of multiple microbial genomes in parallel, making genomic data acquisition quite commonplace nowadays. Sanger sequencing provides read lengths of approximately 1,000 bp/read at a cost of at least $500/Mbp, and now, some NGS technologies, such as platforms from Illumina, Dover Systems, and Life Technologies, can rapidly produce at least 13–36 bps/read for well under $1.00/Mbp (Shendure and Ji 2008). Due to the plummeting costs of using these high-throughput platforms for genome sequencing, several large-scale sequencing initiatives have commenced, such as the Genomic Encyclopedia of Bacteria and Archaea (Wu et al. 2009), the Broad Institute's initiative to sequence 20 actinomycete genomes, the *Sorcerer II* Global Ocean Sampling sequencing effort (Biers et al. 2009), and the Human Microbiome Project (Consortium 2010). In addition, the number of complete whole microbial genome sequences has skyrocketed, providing biologists and biochemists with high volumes of sequence data. Ultimately, these data have led to a surge in microbial genome mining by expanding the tools available for natural product drug discovery.

Over the last few years, Fenical and coworkers at the Scripps Institution of Oceanography have identified some metabolites produced by microbial gene clusters using specific fermentation conditions, and now, with access to genomic information, a combination of cloning, heterologous gene expression, and fermentation is now being used to "unlock" these metabolites. For example, using a reverse or "bottom-up" genetic approach, Udwary et al. sequenced *S. tropica*, the first actinomycete to require seawater for fermentation, and determined that it dedicates a significant percentage of its genome to natural product biosynthesis (~10 %) (Udwary et al. 2007). The authors identified 17 potentially novel biosynthetic gene clusters, including the salinosporamide locus. Bioinformatic analysis of the genome of *S. tropica* strain CNB-440 led to the identification of the *slm* cluster, a cryptic 10-module PKS gene cluster, which led to the identification and structural elucidation of a series of polyene macrolactam polyketides lacking a sugar or other additional functionalities. Upon inspecting the fermentation broth of *S. tropica* CNB-440 for molecules with the UV characteristics of polyenes, the novel antibiotic salinilactam **44** (Fig. 11) was isolated and determined to have a macrolactam framework matching the predicted metabolic product of the *slm* gene cluster. With NGS technologies, many more reports on the (meta)genomic sequencing of microorganisms and microbial consortia will be published, as scientists have realized there are numerous gene clusters dedicated to secondary metabolism that remain to be known or activated/expressed. See the latest reviews by Ochi and Hosaka for an in-depth analysis of the methods used to sequence microbial genomes (singular as well as a consortium), activate cryptic gene clusters, and isolate natural products (Ochi and Hosaka 2013).

3.1 Using Metagenomics to Identify New Bioactive Molecules from Unculturable Microbes

Less than 1 % of microorganisms are culturable under standard laboratory conditions, making the uncultured majority an untapped resource of chemical diversity. Unculturable eubacteria are proposed to comprise 70 % of all known prokaryotic phyla (Achtman and Wagner 2008), and roughly half of all known natural products with antimicrobial, anticancer, and antiviral activity are produced by prokaryotes, namely filamentous actinomycetes, myxobacteria, cyanobacteria, and members of the genera *Bacillus* and *Pseudomonas* (Berdy 2005). Moreover, 16S rRNA sequencing, fluorescent in situ hybridization (Langer-Safer et al. 1982), and DNA reassociation kinetics (Torsvik et al. 1990, 1998) have estimated that there are 10^5 uncultured eubacterial species per gram of soil and a ubiquitous number of unique uncultured extremophiles that dwell in the marine world and other extreme environments, such as solfataric hot springs, glacier soil, and glacier ice. Thus, a large proportion of prokaryotes remain to be fully exploited for their capacity to produce bioactive metabolites. Filamentous fungi are also prolific producers of secondary metabolites, as genomic sequencing of some of the best studied *Aspergilli* has more secondary metabolites to be discovered. Thus, the microbiomes of environmental samples composed of fungi and prokaryotes represent a large reservoir of unexplored biodiversity. Moreover, researchers hypothesize that many unique metabolites are produced within the context of symbiosis, increasing the potential value of studying these extremophiles to find new bioactive agents.

Metagenomic sequencing or metagenomics, a term coined by Handelsman et al. (1998), has provided new approaches for teasing out genomic sequence data from a community of organisms to identify genes involved in the biosynthesis of new compounds. Advances in metagenomics have helped researchers circumvent the traditional barriers to natural product discovery using molecular methods. Metagenomic sequence data can reveal the true capabilities of unculturable microbes to produce secondary metabolites and has enabled the identification of new compounds independent of our ability to culture these microbes. Thus, metagenomics has become a very powerful tool for identifying gene clusters from environmental samples of mixed, unculturable organisms, providing further insight into their potential to produce secondary metabolites as well as the interactions or symbiosis between a mixed population of organisms (von Mering et al. 2007). Importantly, these culture-independent approaches can assess the metabolic and taxonomic diversity within microbial communities to find new bioactive agents, complementing or, in some cases, replacing culture-based approaches to harvesting new natural products. Sequence- and function-based metagenomic methods have been developed to uncover the metabolic diversity of environmental samples and identify new biosynthetic clusters of interest. Both methods involve cloning environmental DNA and constructing metagenomic libraries that are subsequently transformed in a host. Refer to Sect. 2.2 for ways in which genomic libraries can be heterologously expressed.

3.1.1 Sequence-Based Metagenomic Analyses

Sequence-based metagenome mining involves the use of DNA probes, such as PCR-based sequence tags constructed from conserved regions of known genes or protein families, to identify genes and homologous sequences involved in natural product biosynthesis from genomes that could not otherwise be assembled. This method is useful for searching well-studied classes of genes from random pools of DNA. However, the sequence similarities and repetitive nature of biosynthetic gene clusters can make this method for finding new natural products challenging. Examples of the successful use of sequence-based metagenome mining to identify new secondary metabolites have been reported by Donia et al. as well as the Brady and Jaspars research groups (Donia et al. 2011b; Milshteyn et al. 2014 and Long et al. 2005).

Donia and coworkers used metagenomic sequencing to identify gene clusters corresponding to the production of cyanobactin ribosomal peptides from environmental DNA extracted from dotting colonies of *Didemnum molle* ascidians collected from the tropical Pacific (Donia et al. 2011b). Using unpurified DNA isolated from small colonies, a PCR-based approach using primers based on conserved cyanobactin sequences was used to recover gene clusters involved in producing novel natural products. A gene cluster encoding the biosynthesis of a new ribosomal peptide minimide **45** (Fig. 11) was identified and later cloned and heterologously expressed it in *E. coli* using a dual-vector system. *E. coli* can now be used to produce the marine invertebrate-derived minimide **45** for preclinical testing, as a titer of only <10 μg of minimide per ascidian colony was previously obtained. Although several steps were required to optimize the transcription machinery of the *E. coli* host and dual-vector system used, this is a great example of heterologous expression of a marine-derived recombinant gene cluster in *E. coli*.

Another major hurdle in the development of drugs based on marine-derived compounds, particularly those from marine sponges and ascidians, has been identifying the producing organisms and replenishing their supply. To this end, sequence-based metagenomic analyses can also be used to identify unprecedented biosynthetic gene clusters in a community of unculturable microorganisms that produces novel marine natural products. For example, Haywood and coworkers were able to identify a putative bryostatin 1 **46** (Fig. 11) PKS gene cluster, *bryA*. Bryostatin 1 is a marine-derived anticancer which inhibits protein kinase C. Since protein kinase C is also involved in learning and memory, it may also have potential as an anti-Alzheimer's agent. Obtaining a sufficient supply of this metabolite has limited its clinical development. To find the genes involved in its biosynthesis to obtain more bryostatin 1 **46**, a 300-bp fragment of a β-ketoacyl synthase involved in the biosynthesis of bryostatin was used as a probe to screen a metagenomic library enriched with the DNA of the eubacterial symbiont of the bryozoan *Bugula neritina*, *Candidatus Endobugula sertula* (Hildebrand et al. 2004; Sudek et al. 2006). The *bryA* gene product that was found may be useful in the semi-synthesis of bryostatin 1 **46** (Fig. 11) and analogs as well as finding other genes involved in bryostatin 1 **46** biosynthesis.

The use of PCR-based "sequence tags" also facilitates the mining of metagenomes of any complexity (e.g., highly complex metagenomes with 10^4–10^5 unique species of organisms) by exploiting the fact that most natural products belong to a small number of specific classes (e.g., non-ribosomal peptides, polyketides, alkaloids, ribosomal peptides, terpenes, and sugars) with highly conserved biosynthetic genes. Sequence tags with no close relatives based on bioinformatics analyses using platforms, such as Environmental Surveyor of Natural Product Diversity (eSNaPD) (Reddy et al. 2014) and Natural Product Domain Seeker (NaPDOS) (Ziemert et al. 2012), most likely encode unique gene clusters that produce novel metabolites. Using PCR amplicons with bar-coded natural product sequence tags, the Brady research group identified previously undetected gene clusters involved in the biosynthesis of biomedically relevant molecules from the metagenome of a soil sample collected in New Mexico (Owen et al. 2013). They also created a pipeline to identify gene clusters involved in producing molecules that are structurally related to clinically relevant molecules from environmental samples and may have improved properties, such as enhanced potency, solubility, and bioavailability. These tools will be useful for target identification for drug discovery pipelines as well as the cataloging of natural product biosynthetic gene clusters along with their geographical distribution.

A combination of PCR-based targeting and pyrosequencing, in which many short reads are obtained per amplicon, has also been used to gain more insight into the chemical diversity within a microbial community. Iwai and coworkers termed this strategy gene-targeted metagenomics (GT-metagenomics), in which sequence information can be used to design probes to recover full-length genes from clades of interest (Iwai et al. 2009). Using GT-metagenomics, Iwai and coworkers were able to identify aromatic dioxygenase genes in polychlorinated biphenyl-contaminated soil collected in the Czech Republic using primer sets targeting a 524-bp region conferring substrate specificity in this class of enzymes. Based on sequence alignments of the PCR products, 22 novel gene clusters were detected, excluding known sequences. The identification of these diverse, functional dioxygenase genes provided further insight into genes involved in environmental processes as well as their functional differences and evolutionary origins. This strategy has also been used to determine the diversity of bacterial genes involved in the conversion of ketones into anti-Prelog chiral alcohols in soil samples collected in Japan (Itoh et al. 2014). Although sequence-based approaches can be powerful tools for finding new biocatalysts involved in the production of more diverse chemical scaffolds, the major drawback is that only genes related to known enzymes can be identified.

3.1.2 Function-Based Metagenomic Analyses

Function-based metagenome mining uses identifiable phenotypic traits (e.g., antibiosis, morphological changes, enzyme activity, and pigmentation) commonly associated with secondary metabolite production to identify a clone genome that is already biosynthetically active in a heterologous host. Once an appropriate screening assay is chosen to identify a phenotype of interest, new classes of genes encoding unknown functions can be discovered without requiring any sequence information of the corresponding biosynthetic gene cluster. Notably, function-based screening allows the detection of new genes that are not similar to known genes encoding a specific function.

One of the simplest function-based metagenomic library screens to find new bioactive agents in *E. coli* is an agar-based growth inhibition assay. Brady and Clardy were able to isolate the antibiotic palmitoylputrescine **47** (Fig. 11), previously unreported from cultured organisms, by heterologously expressing DNA isolated from Costa Rican bromeliad tank water in *E. coli*-cosmid libraries (Brady and Clardy 2004). Using a *Bacillus subtilis* overlay assay, an antibacterial clone determined to produce palmitoylputrescine **47** was identified. Long and coworkers also created a metagenomic bacterial artificial chromosome/clone (BAC) library from eubacterial DNA obtained from the ascidian *Lissoclinum patella* and expressed fragments of *Prochloron* sp. genes in *E. coli* because the native producer of the cytotoxic patellamides is an intractable photosynthetic endosymbiont of the genus *Prochloron*. The authors identified functional recombinant clones that produced cytotoxic patellamide D **48** (Fig. 11) using mass spectrometry (MS) and optimized their heterologous expression to produce this compound at levels of 80–100 ng/ml (Long et al. 2005). Additionally, using a pigment-based agar assay, He and coworkers functionally screened a 250,000-fosmid library in *E. coli* constructed from the DNA of the Japanese marine sponge *Discodermia calyx* to identify new bioactive compounds (He et al. 2012). The functional screen of the *D. calyx* metagenome led to the identification of the major red pigment Zn-coproporphyrin III **49** as well as coproporphyrin III **50**, Zn-protoporphyrin IX **51**, and protoporphyrin IX **52** (Fig. 11).

The area of functional metagenomics is moving toward developing high-throughput platforms for screening environmental DNA to identify, capture, and express cryptic clusters to identify new secondary metabolites. However, there are still challenges to function-based metagenomics related to heterologous expression, the sizes of the captured environmental DNA, the enrichment of metagenomic libraries with the DNA of interest, the toxic genes that may be encoded within the DNA, and the low detection frequency or high hit rate due to several activities being detected in function-based screens (Penesyan et al. 2013). Other heterologous hosts, such as *Agrobacterium tumefaciens*, *P. putida*, and *Burkholderia graminis*, and vectors can be used to optimize gene expression (Milshteyn et al. 2014); however, the trial and error required can make this method excessively time-consuming and labor-intensive. As sequencing costs decrease, a mixture of both sequence- and function-based metagenomics will most likely be used in the future to assess the genetic, enzymatic, and chemical diversity within microbial communities.

3.1.3 Exploring Phylogenetic Diversity of Microbial Communities to Find Novel Natural Products

Metagenomic datasets are also used to determine taxonomy and phylogenetic diversity by "binning" sequences of the same phylogeny that are derived from microbial consortia present in soil and marine environments. Ribosomal RNA (rRNA) is typically analyzed in a high-throughput manner directly from environmental samples creating a 16S or 18S rDNA gene clone library followed by PCR amplification with degenerate primers and sequencing to identify existing eubacterial or fungal populations, respectively. Youssef and coworkers have created a list of various phylogenetic markers used in bacterial diversity surveys (Youssef et al. 2014). Alternative phylogenetic markers, such as heat shock protein 70 (Hsp70), *recA*/*radA*, RNA polymerase B, and elongation factors Tu and G, have also been used to determine the taxonomic diversity within a microbial community (Venter et al. 2004). Furthermore, large databases of reference sequences, such as Greengenes (DeSantis et al. 2006), SILVA database (Pruesse et al. 2007), or Ribosomal Database Project II (RDP II) (Cole et al. 2003), have been developed for rRNA-based classification.

Today, the rRNA-gene approach is more frequently being replaced or complemented by whole-genome shotgun sequencing to accurately evaluate the taxonomic composition of an environmental sample, avoiding any inherent bias introduced by amplification using phylogenetic markers and cloning of gene fragments. Venter and coworkers were the first to report the application of whole-genome shotgun sequencing to the environmental DNA from the complex ecosystem of the northwest Sargasso Sea and RDP II to evaluate the diversity and species richness of the genomic content (Venter et al. 2004). Although their PCR-based rRNA analyses using a variety of phylogenetic markers provided a similar qualitative representation of the taxonomical diversity, the quantitative picture varied for different taxonomic groups. The authors successfully showed how direct sequencing can provide a comprehensive and unbiased way to examine species diversity and identify the genetic diversity within gene families. Tyson and coworkers also used a combination of whole-genome shotgun sequencing and PCR-based rRNA analyses to determine the composition and function of simpler microbial communities in samples collected from acid mine drainage biofilm collected in the Richmond mine at Iron Mountain, CA, USA (Tyson et al. 2004). The simple community structure enabled nearly complete genomes of *Leptospirillum* group II and *Ferroplasma* type II organisms to be constructed, providing insight into how each organism depends on each other to obtain carbon or nitrogen and maintains a neutral intracellular pH in a highly acidic, metal-rich environment. Thus, being able to assign functions to all of the DNA fragments can provide models or at least further insight into communities of organisms in unique environments.

Sequence data can also be used to reveal the close associations between organisms, such as tunicates, and uncultivatable microbial symbionts, which fulfill primary and secondary metabolic functions. Using sequenced PCR-amplified 16S rRNA genes, shotgun metagenomic sequencing, confocal microscopy, end sequencing of a fosmid library, and cultivating actinobacteria, Donia and coworkers analyzed the complex microbiome underlying the symbiosis between the marine tunicate *L. patella* and the cyanobacterium *Prochloron didemni*, which produces the cytotoxic patellamides and other therapeutics (Donia et al. 2011a). This study not only determined the nutrient exchange central to the symbiosis between these organisms, but it showed that *L. patella* harbors a variety of other eubacteria (i.e., non-*Prochloron* eubacteria) that contribute to its metabolism and have the potential to produce a plethora of new natural products.

3.2 Single-Cell Genomic Approaches to Identify New Unculturable Microbes with the Potential to Produce Bioactive Metabolites

One advantage to metagenomic sequencing is that it provides sequences for all DNA present in an environmental sample. However, a major drawback to the metagenomic sequencing of complex microbiomes is that biosynthetic genes cannot be correlated to a specific producer, as contigs containing both biosynthetic genes and phylogenetic markers are nonexistent and insufficient sequence depth of binning can preclude the assembly of individual genomes. To address this problem, single-cell genomic approaches have emerged to efficiently and quantitatively correlate the phylogenetic identity of environmental microorganisms to their functional genes and metabalomes at the most fundamental level of biology, reducing metagenomic complexity (Siegl et al. 2011). This strategy can complement metagenomic and culture-based approaches because it can identify key metabolic features, evolutionary history, and mutualistic interactions between the unculturable microbes that dominate an environmental sample.

Once an environmental DNA sample is collected, single-cell isolation is performed using differential centrifugation of a community of microbes, fluorescence-assisted cell sorting, microdissection, microfluidic encapsulation, or micromanipulation devices to individually sort samples enriched with a particular microbe into 96-well plates (Piel 2011). After cell lysis (e.g., an alkaline solution, detergent, or heat) and multiple displacement DNA amplification to sequence the genomic DNA of a single microorganism, DNA contigs are obtained. PCR using specific gene primers can then be used to screen the microorganism or unamplified single DNA fragment in each well. Whole-genome amplification can also be performed using bacteriophage Φ29 polymerase to identify specific genes in a genome. However, one challenge that can be faced is DNA contamination; therefore, all reagents and equipment used should be properly decontaminated (Stepanauskas 2012).

These techniques are increasingly being used in combination with metagenomic sequencing to identify new unculturable prokaryotic species and isolate their biosynthetic genes, such as those involved in the biosynthesis of the antitumor agent apratoxin A **53** (Fig. 11) from the intractable mixture of the cyanobacterium *Lyngbya bouillonii* associated with heterotrophic eubacteria (Grindberg et al. 2011). Both metagenomic sequencing and single-celled genomics have also been used to identify the true producers of bioactive metabolites isolated from sponges. Initially, metabolites, such as the polyketide onnamide A **54** (Fig. 11) and the 48 residue ribosomal peptide polytheonamides, were thought to be solely produced by the sponge; however, challenges were encountered when trying to reisolate these bioactive molecules. To address this challenge, Wilson and coworkers used single-celled analysis to examine the microbial consortium associated with *Theonella swinhoei*-type yellow (Y) (Wilson et al. 2014). The authors used primers specific for the genes involved in the biosynthesis of onnamide-type polyketides as well as ribosomal peptides belonging to the polytheonamide structural class to amplify their corresponding genes. Roughly half of the wells containing amplified genes were positive for *Entotheonella* phylotypes based on 16S rRNA analysis. Metagenomic sequencing was used to assemble the genome of the unculturable bacterium, and two chemically distinct *Entotheonella* sp. (TSY1 and TSY2) were identified with large genome sizes of >9 Mb, which are the largest prokaryotic genomes currently reported. Surprisingly, a number of gene clusters involved in polyketide and ribosomal or non-ribosomal peptide biosynthesis were identified, mainly from *Entotheonella* sp. TSY1, indicating that the *Entotheonella* species are rich sources of structurally diverse metabolites. The 16S rRNA sequences of the two species were 82 % identical to known eubacterial phyla. Thus, these eubacteria were deemed to be a part of the new phylum *Tectomicrobia*, which apparently are widespread among 37 taxonomically diverse sponge species from 20 different locations. Interestingly, the sponges have non-overlapping metabolic profiles based on the location from which *T. swinhoei* Y was collected. Of the 37 taxonomically different sponges, 28 of them were found to harbor species belonging to the genus *Entotheonella,* providing new, abundant sources for drug discovery.

Marine sponges are an important source of bioactive natural products, and the supply of these metabolites has often precluded their evaluation in clinical trials. For years, researchers speculated that symbionts were involved in the production of these metabolites. Now, molecular tools are now revealing the true natural product producers from chemically rich sponges and other marine invertebrates, revealing the significant roles uncultured prokaryotes played in the chemistry of their hosts. With tools to identify the actual producers of these metabolites, researchers can now determine ways to cultivate these organisms or heterologously express or reengineer the genes involved in biosynthesis. Molecular techniques have also shed light on how these microorganisms can be cultured under laboratory settings. For example, metagenomic sequencing was used to direct the microbial culturing conditions by uncovering the genes' presence in microbial communities, such as the microbiota of the medicinal leech *Hirudo verbena,* which contains *Aeromonas veronii* and a *Rikenella*-like eubacterium. High expression levels of mucin and glycan utilization genes were found in the *Rikenella*-like eubacterium inside the crop, and growing the microbe in media containing mucin instead of glucose led to the growth of pure cultures of this microorganism (Bomar et al. 2011). These methods can also be used to determine the appropriate conditions for culturing microorganisms to produce new secondary metabolites.

With the increasing availability of metagenomic sequence data facilitating more "bottom-up" approaches to drug discovery, the number of bioactive agents listed in Table 2 will continue to increase exponentially. In addition, more tools are being developed to facilitate function—(e.g., enriching metagenomic libraries and cloning of environmental DNA) and sequence-based metagenome mining (e.g., improvements in DNA cloning), increasing the number of ways researchers can access novel natural products without having to rely on culture-dependent methods. Importantly, the integration of single cell with bioprospecting new gene

Table 2 Bioactive compounds isolated from reengineered microorganisms

Microorganism	Compound	Bioactivity	References
Salinispora tropica CNB-440	Salinilactam **44**	Antimicrobial	Udwary et al. (2007)
Didemnin molle; heterologous expression in *E. coli*	Minimide **45**	N/A	Donia et al. (2011a, b)
Possibly the unculturable microbe *Candidatus Endobugula sertula* associated with *Bugula neritina*	Bryostatin **46**	Antiviral; inhibitor of protein kinase C isozymes; cytotoxic	Newman (2012)
Heterologous expression of bromeliad water tank environmental DNA in *E. coli*	Palmitoylputrescine **47**	Antimicrobial	Brady and Clardy (2004)
Prochloron sp. associated with *Lissoclinum patella*	Patellamide D **48**	Cytotoxic	Degnan et al. (1989), Long et al. (2005)
Heterologous expression of *Discodermia calyx* metagenome in *E. coli*	Zn-coproporphyrin III **49**	N/A	He et al. (2012)
Heterologous expression of *Discodermia calyx* metagenome in *E. coli*	Coproporphyrin III **50**	N/A	He et al. (2012)
Heterologous expression of *Discodermia calyx* metagenome in *E. coli*	Zn-protoporphyrin IX **51**	N/A	He et al. (2012)
Heterologous expression of *Discodermia calyx* metagenome in *E. coli*	Protoporphyrin IX **52**	N/A	He et al. (2012)
Lyngbya boiullonii; *L. majuscule*; *L. sordida*; and *Moorea producens*	Apratoxin A **53**	Cytotoxic	Luesch et al. (2001)
Unknown eubacteria from *T. swinhoei* and *Theonella* sp.	Onnamide A **54**	Antiviral; cytotoxic	Sakemi et al. (1988), Burres and Clement (1989)

N/A information not available

clusters involved in secondary metabolism holds significant potential for finding new drug candidates. For more details on using metagenomics to exploit uncultivated eubacteria, see the latest 2013 review by Wilson and Piel (2013) and 2014 reviews from the Brady group (Charlop-Powers et al. 2014; Milshteyn et al. 2014).

4 Dereplication Methods for Identifying New Secondary Metabolites from Extremophiles

Some of the reasons most major biopharmaceutical companies phased out their natural product drug discovery programs in the 1990s were related to high rediscovery rates and limited technology required to isolate and characterize chemically diverse metabolites from highly complex mixtures. This time happened to coincide with the extensive development of high-throughput screening technology and combinatorial chemistry, which were used to synthesize libraries of simpler, more "drug-like" molecules against defined molecular targets. However, from 1981 to 2010, the number of approved drugs that are or derived from functionally diverse, structurally complex natural products has continued to supercede the one de novo combinatorial compound that has been approved as a drug in this 30-year time frame (Newman and Cragg 2012). As a result, pharmaceutical companies developed smaller, more targeted synthetic libraries based on natural product pharmacophores rather than relying on the mindless assembly of molecules produced by combinatorial chemistry (Atuegbu et al. 1996; Xu et al. 1999). This shift in focus may have also renewed the interest in natural product drug discovery and finding pharmacophores that can be improved using diversity-oriented synthesis.

In addition to finding natural products with unique bioactivity, the limitations to working with these complex compounds have been the following: (1) Building a high-quality natural products library is not trivial; (2) natural products are found in minute quantities in extracts; (3) often times, there are high rediscovery rates of these metabolites; (4) dereplication is very time-consuming, labor-intensive, and expensive; and (5) the structural complexity of natural products often precludes their structural validation and derivatization by chemical synthesis. To address these bottlenecks in natural product drug discovery, academic and some industrial natural product chemists have been working on developing more streamlined approaches for the production and analysis of natural products for drug discovery.

The methods used to prepare extracts from terrestrial and marine sources as well as perform bioassay-guided fractionation and structural elucidation have become more automated (Eldridge et al. 2002; Johnson et al. 2011). Furthermore, with the increasing amounts of genomic sequence data, there is now a greater push to develop new analytical instrumentation tightly integrated with bioinformatics and computational design resources to provide correlations between genes, enzymes, and chemical structures in a refined, high-throughput manner. High-field nuclear magnetic resonance (NMR) and MS using a variety of ionization methods have traditionally played a central role in de novo structural determination of secondary metabolites for the past 40 years. MS was later coupled to high-performance liquid chromatography (LC) to dereplicate crude extracts, and this system has been further developed over the years as reported in the recent review by Carter (2014). Today, these analytical tools are being coupled with using principal component analysis (PCA) and bioinformatics to create molecular frameworks, facilitating correlations between genomic and analytical data to connect new chemotypes with genotypes. These advances are now shifting the focus in natural products research from the isolation and identification of single, pure compounds to the identification of mixtures of compounds.

4.1 Hyphenated Liquid Chromatography Techniques for the Identification and Characterization of Natural Products

Several combinations of LC hyphenated techniques have been developed over the last three decades (e.g., LC-UV, LC-MS, LC-MS/MS, LC-ELSD-MS, LC-UV-SPE-NMR, LC-NMR, and LC-MS/MS-NMR) to provide structural information. Advances in chromatography, such as the use of supercritical fluid or different solid-phase extraction (SPE) materials, have improved the LC resolution power required to dereplicate complex extracts and obtain high-quality metabolomics data. In addition, the electrospray ionization (ESI) method enabled the generation of ions from natural product extracts comprising polar, nonvolatile compounds. With LC-MS and LC-NMR techniques, molecules can be identified without isolating pure amounts of the individual components of a mixture. Using LC-UV-MS systems, the acquired molecular mass and UV absorption data of a compound in an extract can be correlated with known compounds in natural product databases, such as the Dictionary of Natural Products, Antibase, and Marinlit, to identify compounds. Liquid chromatography–high-resolution mass spectrometry (LC-HRMS) is a useful, sensitive technique for dereplicating extracts and structurally characterizing individual components based on mass fragmentation patterns, whereas LC-NMR is a less sensitive technique that enables full structural characterization of each component, with the exception of silent functional groups (i.e., hydroxyl and amino moieties) due to exchangeable hydrogen atoms (Halabalaki et al. 2014). When the absolute configuration of an isolated molecule cannot be determined, synthesis and single-crystal X-ray crystallography are still utilized. Although these hyphenated techniques have created major breakthroughs in the rapid identification of natural products, large datasets obtained from untargeted metabolomics can be overwhelming to sort through, especially when chemically complex extracts contain low amounts of a unique natural product and large quantities of known compounds.

Using hierarchical cluster analysis (HCA) and PCA, data points within a large, complex 3-dimensional dataset can be clustered and reduced via multivariate analysis to discriminate among samples, and large libraries of microbial extracts can be dereplicated in a single analysis (Forner et al. 2013). Specifically, HCA creates a hierarchy of clusters based on different algorithms and distance definitions, whereas PCA efficiently uses multivariate analysis to determine the variance within a dataset, which can be visualized in a 3-dimensional plot. By integrating LC-HRMS, PCA, and HCA, Forner and coworkers showed that they can reproducibly group eubacteria based on their metabolic profiles and discriminate between closely related strains of microorganisms, such as the taxonomically indistinguishable *Streptomyces* (Forner et al. 2013). Twenty-two strains of closely related *Streptomyces* collected from the sites off the Atlantic Canadian coast and the Bahamas were selected and taxonomically characterized by 16S rDNA sequencing. These strains were fermented in triplicate, extracted with ethyl acetate, and the corresponding extracts were chemically dereplicated, clustered according to their chemical footprint or bar-coded LC-HRMS data, and metabolites responsible for differences between chemical footprints among different microbial strains were highlighted by PCA. To validate this approach, this method was then applied to a larger set of 120 isolates obtained from three sites off the coast of Newfoundland, Canada, which were also chemically dereplicated and clustered into 37 clades, revealing putatively novel metabolites with unique *m/z* ratios [i.e., $[M + H]^+$ 1038.6278 and $[M + Na]^+$ 1060.6089]. Not only can these techniques be used to find new metabolites, but they can also be used as a large-scale chemotaxonomical tool to correlate the metabolic footprints of microbial extracts determined by untargeted metabolomics to a specific genus and/or species.

Using a similar method without HCA, Cortina and coworkers performed a differential analysis of the molecular features of extracts prepared from wild-type and targeted mutant libraries of *M. xanthus* DK1622. The strains were grown in quadruplicate and analyzed by LC-HRMS, and the data obtained were treated using a compound finding algorithm, ultimately resulting in the identification of >1,000 molecular features per sample. The chemical features missing in the extracts prepared from mutants were identified by PCA of the preprocessed LC-HRMS datasets. Notably, three hidden, low abundant metabolites were found, namely a new non-ribosomal peptide and polyketide sugar, as well as the complete characterization of the NRPS-PKS hybrid myxoprincomide **55** (Fig. 12)

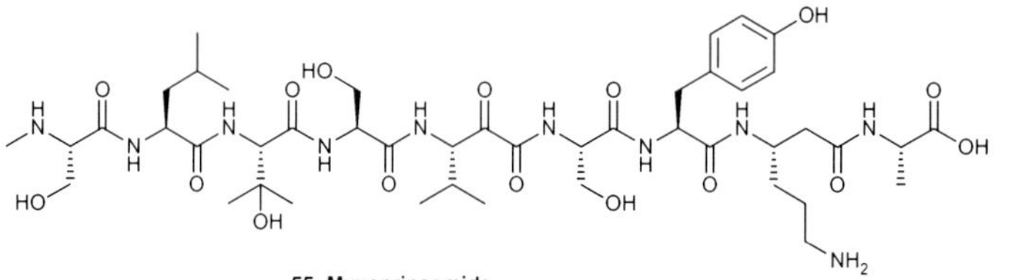

55. Myxoprincomide

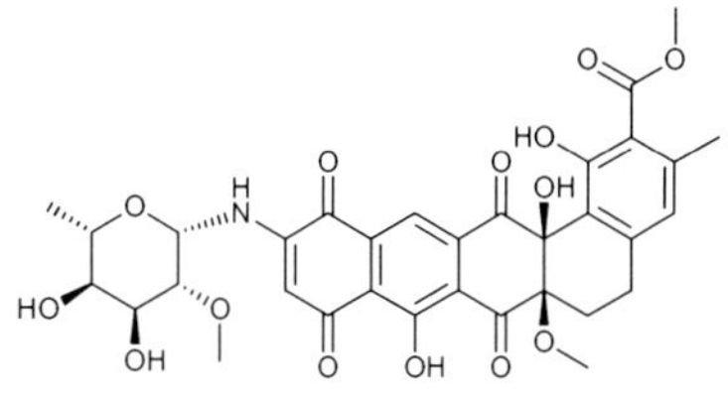

56. Arenimycin A

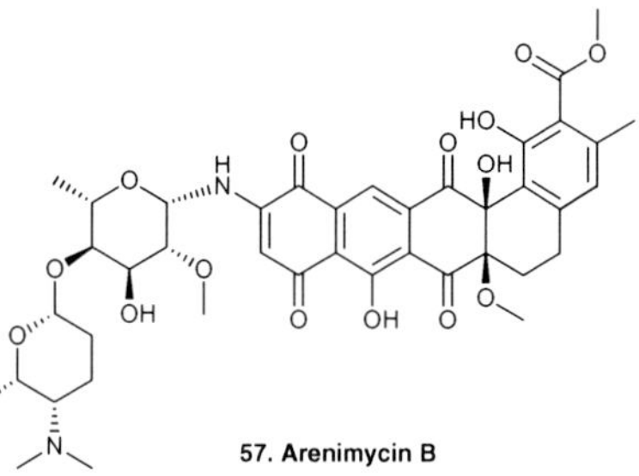

57. Arenimycin B

Fig. 12 Structures **55–57**

(Cortina et al. 2012). Using *M. xanthus* DK1622 genome sequence data, these metabolites were correlated with their producing gene clusters. These results demonstrate how improving the resolution for discriminating among the chemical diversity of a variety of microbial strains can facilitate the identification of new metabolites and assignment of their corresponding biosynthetic gene clusters of the genomic information.

4.2 Soft Ionization Mass Spectrometry Methods Used the Identification and Characterization of Natural Products

Advances in MS have enabled high-throughput natural product discovery, determining correlations between chemotype, genotype, and species, a better understanding of interspecies interactions, as well as a way to aid in identifying the bioactivity of novel compounds. However, the bioactivity of natural products is still commonly determined using target-mediated inhibition. Determining the true biological function of these small molecules within the context of a microorganism's ecosystem has been limited by the lack of in situ technology for real-time monitoring of the in vivo production of natural products. Over the past two decades, other methods of generating ions, such as desorption ionization methods, have been developed to generate mass ions from thermally labile and nonvolatile solids. One of the most important desorption techniques is MALDI-assisted laser desorption/ionization (MALDI), which has been used to obtain MS data of peptides and proteins (>1,000 Da), and MALDI devices have been developed to be simple, have high throughput, be inexpensive, and be efficient such that they are now used in clinical settings for epidemiological studies and taxonomical classification of microorganisms.

Low molecular weight natural products can be identified in real time using various desorption and softer ionization techniques without requiring sample pretreatment. Imaging mass spectrometry (IMS) is a tool that has emerged for natural product scientists to get a 2-dimensional snapshot of the distribution of mass ions that correlate with natural products produced by a growing colony of intact cells over time. Various desorption and softer ionization methods have been applied to IMS in order to profile new bioactive small molecules (<500 Da) on biological surfaces. IMS increases the likelihood of finding unique ion distribution patterns and molecular phenotypes that cannot be observed with the naked eye. Previously unidentified metabolites, such as brominated pyrrole-2-amino imidazole alkaloids produced by the sponge *Stylissa flabellata* (Yarnold et al. 2012), have been identified using this technique that were not detected in the LC-MS analysis of extracts due to ion suppression and matrix interferences. Chemotypes, such as anti-infective agents, can also be linked to antagonistic phenotypes of a target microorganism against a pathogen based on the formation of inhibition zones. In addition, the localization of molecules can be rapidly visualized over time, providing insight into their ecological functions. Using IMS, Yarnold and coworkers were able to visualize the localization of metabolites in three chambers (i.e., pinacoderm, mesohyl, or choanoderm) of *S. flabellata* and speculate about their function and biosynthesis. For more on IMS, see the recent 2014 reviews by Shih et al. (2014) as well as the Dorrestein research group (Bouslimani et al. 2014).

Other efficient methods to profile the metabolomes of extremophiles, such as marine cyanobacteria, have also been developed to connect phenotypes to chemotypes. Sensitive detection techniques, such as MS/MS, have helped sort through complex datasets and identify and quantify a specific compound in an extract. The Dorrestein research group at The University of California San Diego and Scripps Institution of Oceanography recently developed MALDI-based IMS (Watrous and Dorrestein 2011) to identify, quantify, and visualize the metabolome and metabolic exchanges directly from individual microorganisms and marine assemblages, which comprise multiple metabolomes (Yang et al. 2011).

Extending MALDI-based IMS, Kersten and coworkers developed a workflow that combines bioinformatics and de novo tandem mass spectrometry (MS^n) to connect chemotypes of peptide natural products (i.e., ribosomal and non-ribosomal peptide natural products) detected from agar plates to their corresponding genotypes (Kersten et al. 2011). Using MS^n data processing software, partial peptide structures are assembled to produce core peptide sequences that can be correlated with natural product peptide biosynthetic pathways. This is also known as peptidogenomics. The authors conducted a series of proof of concept experiments using known peptides and identified 14 new natural product peptides from different sequenced strains of *Streptomyces*. Extending MS-guided genome mining, Kersten and coworkers also used this technique to identify new glycosylated natural product chemotypes. Oligosaccharides, such as glycans, were sequenced based on the cleavage of *O-/N*-glycosidic bonds, and MS^n fragments corresponding to arenimycins A and B **(56–57)** (Fig. 12) were identified from a strain of *Salinispora arenicola* CNB-527 (Kersten et al. 2013b). Thus, MS-guided genome mining is a useful way of connecting metabolomes to glycogenetic codes, which can be further exploited to produce increased amounts of a particular metabolite via heterologous gene expression or metabolic engineering.

4.3 Developing Molecular Frameworks to Find New Pharmacophores

With the addition of the new molecular networking software Tandem-MS Origin Tracing Engine (TOrTE), predicted structures obtained from MS^n fragmentation patterns of original extracts and pure "seed" compounds for each compound in a molecular network found in extracts can be easily visualized, connecting chemotaxonomy, expression analysis, biosynthetic mechanisms, and microbial cultures (Winnikoff et al. 2014). Differences between the MS^n spectra of metabolites can be computed and used to determine structural similarity. The Dorrenstein and Gerwick research groups applied this method to a cyanobacterial culture collection, which included strains that produced known metabolites and their corresponding analogs. Several producers of known metabolites, such as malyngamide C, with notable bioactivity were identified, demonstrating that this method can be used to identify more abundant sources of metabolites of interest.

Recently, Doroghazi and coworkers developed a global classification framework for natural product biosynthetic gene clusters (e.g., those encoding NRPSs, PKSs, NRPS-independent siderophores, lantipeptides, and thiazole–oxazole-modified microcins) and analyzed biosynthetic pathways in 830 actinomycete genome sequences, including 344 of which were sequenced for this study (Doroghazi et al. 2014). The distance between the numbers of homologous genes, nucleotides involved in the pairwise alignment, and sequence identity of the amino acids within the domains of modules between two natural product gene clusters was determined to group unrelated gene clusters. This framework is advantageous for analyzing gene clusters composed of large genes with highly repetitive sequences (e.g., those encoding PKSs and NRPSs) and have very similar functions. The network created by the authors, comprising 11,422 gene clusters, which essentially collapsed into 4,122 gene clusters based upon similarity, was validated in hundreds of microbial strains by correlating the detection of small molecules by MS with the presence or absence of gene clusters. Using LC-HRMS to test this framework, from 178 extracts prepared from different strains grown in different fermentation media, 2,521 unique compounds were identified and 27 known structural families were eventually linked to nine biosynthetic gene clusters. This framework identified the correct gene cluster in 92 % of the genomes of the correct source organism and revealed that actinobacteria have the capacity to produce thousands of new drug leads. Based on these results, the authors are now considering imposing more constraints on the correlations by coupling MS^n data with genomic data to improve gene cluster networks.

4.4 NMR Techniques for the Structural Characterization of Low Concentrations of Pure or Mixtures of Compounds

New NMR techniques have recently been developed to successfully elucidate complex chemical structures, in which the ratio of hydrogen atoms to oxygen, sulfur, or nitrogen atoms is less than two, suggesting that the Crews rule may need to be revised (Senior et al. 2013). Usually, a combination of two-dimensional NMR experiments, such as heteronuclear bond correlation (HMBC) and 1,1-adequate double quantum transfer experiment (ADEQUATE), is used to distinguish $^2J_{CH}$ from $^3J_{CH}$ correlations. However, this becomes problematic with proton-deficient natural products, in which $^4J_{CH}$ correlations are required to establish connectivity between carbon atoms. To address this problem, a variation of the 1,n-ADEQUATE experiment, the dual-optimized inverted $^1J_{CC}$

1-n-ADEQUATE experiment, was developed to provide $^1J_{CC}$ correlations with high sensitivity (Senior et al. 2013). In this experiment, two pairs of $^1J_{CC}/^nJ_{CC}$ optimizations are used to obtain broadband (29–82 Hz) inversion of $^1J_{CC}$ correlations in the 1,n-ADEQUATE spectrum, distinguishing protons and carbons one bond away from those three bonds away with $^3J_{CC}$ correlations that are otherwise the predominant correlations in the spectrum. Other second-tier experiments have been developed to establish connectivity, some faster than others, between heteronuclear atoms, such as inverted direct response (IDR)-HSQC-TOCSY (Blinov et al. 2006), LR-HSQMBC (Williamson et al. 2014), ASAP-HMQC (Schulze-Sünninghausen et al. 2014), and ASAP-HSQC (Schulze-Sünninghausen et al. 2014), which can be used with non-uniform sampling (roughly 25 %) to save time while collecting a sufficient data on a small amount of sample. In addition, residual dipolar couplings in NMR spectra (Gil 2011) can be used to determine the relative configuration of complex molecules and pure shift NMR experiments (Aguilar et al. 2010, 2011) can aid in improving the resolution of peaks to match that achieved in the signal-to-noise ratio. For more information on other improved techniques to characterize natural products, see the 2013 review by Breton and Reynolds (2013).

Using hyphenated technology in a high-throughput manner to identify new chemical entities from extremophiles will lead to a significant increase in the number of compounds listed in Table 3. Importantly, being able to integrate these technologies with genomic sequence data will lead to the rapid characterization of new metabolites, which can be correlated with natural product extracts associated with a specific genus and/or species. Several advances are being made to analytical technology with respect to algorithms, ion sources, hardware, sensitivity, size, separation speed, and (spacial) resolution, speeding up the time it takes to dereplicate natural product extracts. Furthermore, with the generation of more data, we will see contributions from other fields, such as mathematics, physics, and computer science, to enable rapid data manipulation and statistical analyses as well as develop interactive databases for the large-scale analysis of natural products. The field of genomics has dealt with how to deal with high volumes of data, and although molecular networking is one tool that is starting to address how to deal with the complexity of metabolomics data, other software and algorithms need to be developed to further streamline the drug discovery workflow.

Table 3 Bioactive compounds isolated from reengineered microorganisms

Microorganism	Compound	Bioactivity	References
Mutant strains of *Myxococcus xanthus* DK1622	Myxoprincomide **55**	N/A	Cortina et al. (2012)
Streptomyces arenicola CNB-527	Arenimycins A–B **(56–57)**	Antiviral; cytotoxic	Kersten et al. (2013b)

N/A information not available

5 Summary and Concluding Remarks

The identification, isolation, and structural elucidation of bioactive secondary metabolites are still at the heart of all academic and industrial natural products chemistry research programs, as new therapeutics are needed in contemporary medicine. Today, bioprospecting extremophiles and the development of culture-dependent and culture-independent methods have provided new sources of novel chemical entities. Furthermore, due to the following seven marine-derived natural products, successfully making it through the clinical pipeline, the research community now has a renewed interest in natural product drug discovery: cytarabine **58** (sponge/bacterium), vidarabine **59** (sponge/bacterium), ziconotide **60** (cone snail/mollusk/microbe?), eribulin mesylate **61** (sponge/dinoflagellate?), omega-3-acid ethyl esters (fish/microalgae; e.g., ethyl-eicosapentaenoic acid **62**), ecteinascidin 743 **63** (tunicate/bacterium), and brentuximab vedotin **64** (mollusk/cyanobacterium) (Fig. 13). See Table 4 at the end of the chapter for the bioactivity of each of these drugs. While extremophiles are playing a bigger role in drug discovery, more tools are being developed to identify new pharmacophores.

The information generated by next-generation sequence (NGS) technologies has created a renaissance in natural products drug discovery, enabling researchers to use culture-independent methods to identify new pharmacophores. Initially, these data were used to optimize the production of well-known, clinically important natural products on an industrial scale, but now research efforts are shifting to using these techniques to discover novel, cryptic metabolites. Notably, genomic data are being used as blueprints for metabolic engineering, eliminating different competitive pathways and modifying transcription and translation machinery in the microbial host to increase the production of secondary metabolites. Scientists are now recognizing that the integration of tools, such as bioassays, metagenomics, single-cell analysis, metatranscriptomics, metaproteomics, metabolic engineering, metabolomics, and heterologous gene expression, has become necessary to uncover the biosynthetic potential of uncultivatable microorganisms, streamlining the drug discovery process. For example, bioassay-guided genome mining of two predicted enediyne biosynthetic pathways in *S. tropica* CNB-440 led to the identification of the type II PKS gene cluster encoding the enzymes involved in the production of the cytotoxic lomaiviticin A **65**, C–E **(66–68)** (Fig. 13) (Kersten et al. 2013a).

Now researchers have moved away from using top-down approaches in natural product discovery, such as using MS- and NMR-based high-throughput screening of microbial extracts prepared using fermentation conditions to identify new compounds, to utilizing more genomics-based or "bottom-up" approaches to guide the isolation of novel metabolites from unculturable microbes. Currently, sequencing technologies (e.g., the Next-Generation Sequencing Simulator for Metagenomics, NeSSM) (Jia et al. 2013) are being further developed to quickly obtain longer reads as well as sequence complete genomes from a single cell in environmental DNA samples. Whole-genome methods are also being developed to

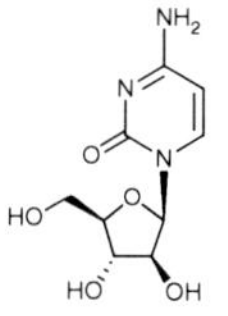

58. Cytarabine

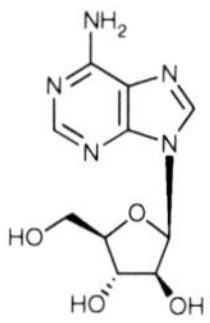

59. Vidarabine

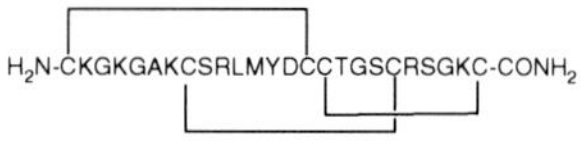
H2N-CKGKGAKCSRLMYDCCTGSCRSGKC-CONH2

60. Ziconotide / SNX-111 / Prialt(R)

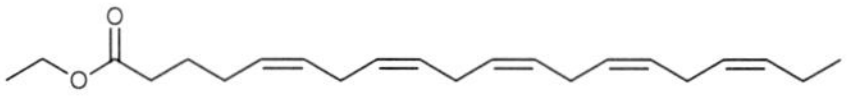

62. Ethyl eicosapentaenoic acid

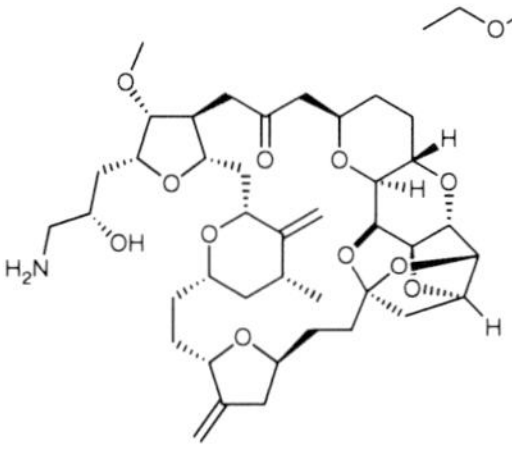

61. E_{7389} (Eribulin)

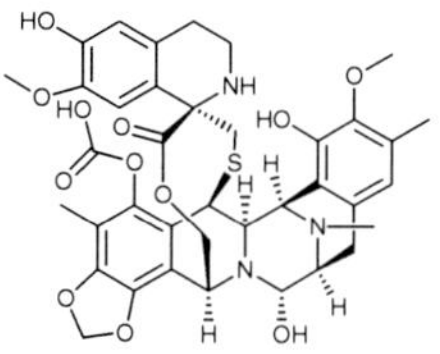

63. Ecteinascidin 743 (Yondelis™)

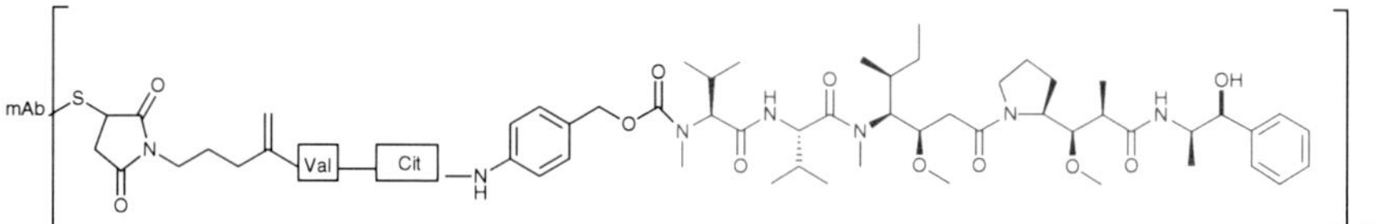
mAb
Val
Cit
8

64. Brentuximab vedotin; Red color is momomethylauristatin E

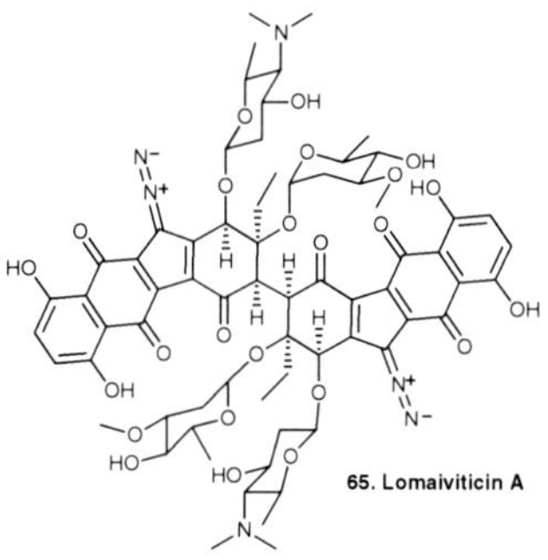

65. Lomaiviticin A

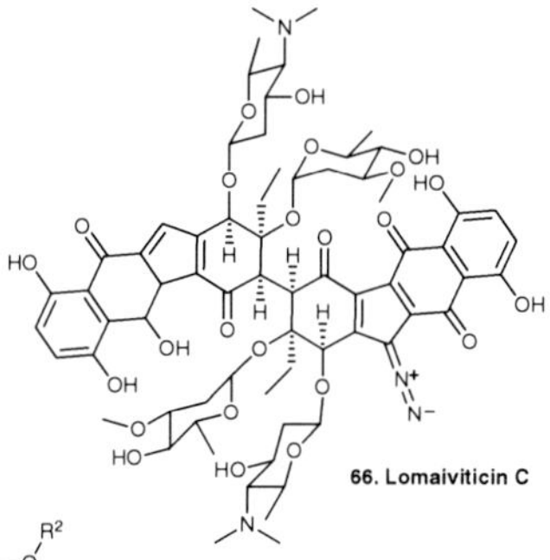

66. Lomaiviticin C

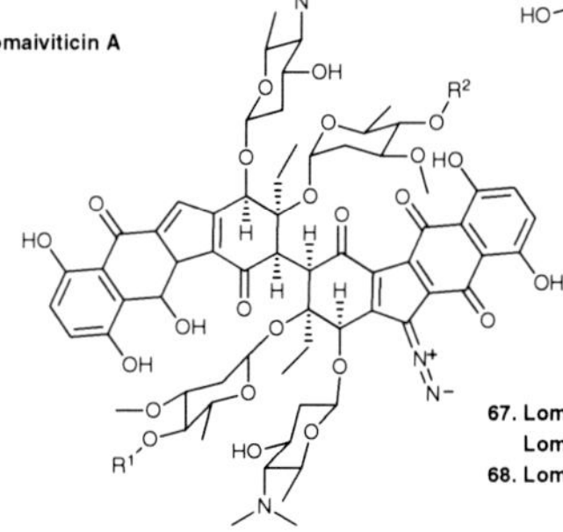

67. Lomaiviticin D; 2 isomers, $R^1 = CH_3$, $R^2 = H$
Lomaiviticin D; 2 isomers, $R^1 = H$, $R^2 = CH_3$
68. Lomaiviticin E; $R^1 = R^2 = CH_3$

Fig. 13 Structures **58–68**

Table 4 Bioactive compounds discussed and their producing microorganisms

Microorganism	Compound	Bioactivity	References
N/A	Cytarabine 58	Cytotoxic	Suckling (1991), Bergmann and Feeney (1950, 1951), Bergmann and Burke (1955)
Streptomyces griseus from Eunicella cavolini	Vidarabine 59	Antiviral; cytotoxic	Cimino et al. (1984)
	Ziconotide 60	N/A	Olivera et al. (1987)
Sponge/dinoflagellate?	Eribulin mesylate 61	Cytotoxic	Kishi et al. (1995)
N/A	Ethyl-eicosapentaenoic acid 62	Antischizophrenic; treatment for bipolar disorder and hypertriglycemia; antidepressant	Kaselein et al. (2014)
Candidatus Endoecteinascidia frumentensis (AY054370) associated with Ecteinascidia turbinata	Trabectedin 63	Cytotoxic	Rath et al. (2011)
N/A	Brentuximab vedotin 64	Cytotoxic	Deng et al. (2013)
Micromonospora LL-37l366 strain isolated from P. lithostrotum; Salinispora pacifica strain DPJ-0019 (NRRL 50168)	Lomaiviticins A 65, C–E (66–67)	Cytotoxic; antimicrobial (65)	He et al. (2001), Woo et al. (2012)

N/A information not available

reduce amplification biases and chimeric DNA rearrangements as well as improve de novo assembly. Importantly, because the amount of data generated by NGS technologies will continue to increase exponentially, new bioinformatics tools are being developed to analyze the taxonomic content of large metagenomic datasets (Metagenome analyzer, MEGAN) (Huson et al. 2007), Kraken (Wood and Salzberg 2014), and Meta-QC-Chain (Zhou et al. 2014) with both complex and repetitive genes and streamline the identification of new pharmacophores and their corresponding gene clusters.

New software is being developed to automate the recognition and classification of biosynthetic gene clusters. For example, community-based Web servers (e.g., antiSMASH, SMURF, NP.searcher, ClustScan, or ClusterMine360) have been developed to facilitate automated searches for genes involved in the production of molecules with specific structures or functions, such as NRPS and/or PKS gene clusters (Conway and Boddy 2013; Caboche et al. 2008; Ichikawa et al. 2013; Blin et al. 2013) to perform in silico predictions (Marques et al. 2013). Genomics-based natural product discovery is providing greater insight into diverging pathways from genomic loci, providing a blueprint for finding new, structurally diverse pharmacophores. New pattern-based, versatile algorithms will need to be further developed to better identify unknown open reading frames as well as predict molecular structures of pure compounds or components in complex mixtures. The global classification of all biosynthetic gene clusters is difficult due to the repetitive features of different classes of enzymes involved in secondary metabolism. However, to address this problem, new systematic bioinformatics frameworks are being developed to globally classify biosynthetic gene clusters into families, revealing the wealth of natural products that exists among microorganisms for drug discovery.

The challenges that remain are the development of high-throughput methods for heterologously expressing gene clusters involved in secondary metabolism as well as screening these molecules for various biological activity. Because of the variation in the GC content of DNA and codon usage, a reduction in translation efficiency is often observed when screening DNA libraries. Although host modifications can be made to improve gene expression, eliminate competing pathways, and provide sufficient quantities of natural product precursors, successful engineering significantly depends on DNA sequences. Faster technology is also required for the assembly of large DNA fragments to streamline heterologous gene expression. Lastly, because the bioactivity of most reported natural products is typically identified using at least one or two target-based tests for cytotoxicity and antimicrobial activity, new, inexpensive, target-based, or whole-cell high-throughput methods for screening natural products will be essential for identifying novel bioactive chemical scaffolds and determining their mechanisms of action (Schulze et al. 2013). With the development of new high-throughput/high-content screens (many published in the Journal for Biomolecular Screening) by pharmaceutical companies and major research institutes, such as the Broad Institute, researchers will be able to quickly identify bioactive components in crude extracts and determine their functions.

References

Achtman M, Wagner M (2008) Microbial diversity and the genetic nature of microbial species. Nat Rev Micro 6:431–440

Aguilar JA, Faulkner S, Nilsson M, Morris GA (2010) Pure shift 1H NMR: a resolution of the resolution problem? Angew Chem Int Ed 49:3901–3903

Aguilar JA, Nilsson M, Morris GA (2011) Simple proton spectra from complex spin systems: pure shift NMR spectroscopy using BIRD. Angew Chem Int Ed 123:9890–9891

Atuegbu A, Maclean D, Nguyen C, Gordon EM, Jacobs JW (1996) Combinatorial modification of natural products: preparation of unencoded and encoded libraries of *Rauwolfia alkaloids*. Bioorg Med Chem 4:1097–1106

Ben-Dov E, Kramarsky-Winter E, Kushmaro A (2009) An in situ method for cultivating microorganisms using a double encapsulation technique. FEMS Microbiol Ecol 68:363–371

Bentley HR, Cunningham KG, Spring FS (1951) 509. Cordycepin, a metabolic product from cultures of *Cordyceps militaris* (Linn.) Link. Part II. The structure of cordycepin. J Chem Soc 2302–205

Berdy J (2005) Bioactive microbial metabolites. J Antibiot 58:1–26

Bergmann W, Feeney RJ (1950) Isolation of a new thymine pentoside from sponges. J Am Chem Soc 72:2809–2810

Bergmann W, Feeney RJ (1951) Marine products. XXXII. The nucleosides of sponges. J Org Chem 16:981–987

Bergmann W, Burke DC (1955) Marine products XXXIX. The nucleosides of sponges III. Spongothymidine and spongouridine. J Org Chem 20:1501–1507

Biers EJ, Sun S, Howard EC (2009) Prokaryotic genomes and diversity in surface ocean waters: interrogating the global ocean sampling metagenome. App Environ Microbiol 75:2221–2229

Bilitchenko L, Liu A, Densmore D (2011) Chapter seven—the Eugene language for synthetic biology. In: Voigt C (ed) Methods in enzymology, vol 498. Academic Press, London, pp 153–172

Blin K, Medema MH, Kazempour D, Fischbach MA, Breitling R, Takano E, Weber T (2013) antiSMASH 2.0—a versatile platform for genome mining of secondary metabolite producers. Nucleic Acids Res 41:W204–W212

Blinov KA, Larin NI, Williams AJ, Zell M, Martin GE (2006) Long-range carbon–carbon connectivity via unsymmetrical indirect covariance processing of HSQC and HMBC NMR data. Magn Reson Chem 44:107–109

Bode HB, Bethe B, Höfs R, Zeeck A (2002) Big effects from small changes: possible ways to explore Nature's chemical diversity. ChemBioChem 3:619–627

Bollmann A, Palumbo AV, Lewis K, Epstein SS (2010) Isolation and physiology of bacteria from contaminated subsurface sediments. App Environ Microbiol 76:7413–7419

© The Author(s) 2015   49

L.-A. Giddings and D.J. Newman, *Bioactive Compounds from Extremophiles*, Extremophilic Bacteria, DOI 10.1007/978-3-319-14836-6

Bomar L, Maltz M, Colston S, Graf J (2011) Directed culturing of microorganisms using metatranscriptomics. mBio 2:e00012–00011

Bouslimani A, Sanchez LM, Garg N, Dorrestein PC (2014) Mass spectrometry of natural products: current, emerging and future technologies. Nat Prod Rep 31:718–729

Brady SF, Clardy J (2004) Palmitoylputrescine, an antibiotic isolated from the heterologous expression of DNA extracted from bromeliad tank water. J Nat Prod 67:1283–1286

Breton RC, Reynolds WF (2013) Using NMR to identify and characterize natural products. Nat Prod Rep 30:501–524

Burres NS, Clement JJ (1989) Antitumor activity and mechanism of action of the novel marine natural product mycalamide-A and -B and onnamide. Cancer Res 49:2935–2940

Caboche S, Pupin M, Leclére V, Fontaine A, Jacques P, Kucherov G (2008) NORINE: a database of nonribosomal peptides. Nucleic Acids Res 36:D326–D331

Carter GT (2014) NP/MS since 1970: from the basement to the bench top. Nat Prod Rep 31:711–717

Cassady JM, Chan KK, Floss HG, Leistner E (2004) Recent developments in the maytansinoid antitumor agents. Chem Pharm Bull 52:1–26

Charlop-Powers Z, Milshteyn A, Brady SF (2014) Metagenomic small molecule discovery methods. Curr Opin Microbiol 19:70–75

Chiang Y-M, Szewczyk E, Davidson AD, Keller N, Oakley BR, Wang CCC (2009) A gene cluster containing two fungal polyketide synthases encodes the biosynthetic pathway for a polyketide, asperfuranone, in *Aspergillus nidulans*. J Am Chem Soc 131:2965–2970

Cimino GS, DeRosa S, DeStefano (1984) Antiviral agents from a gorgonian, *Eunicella cavolini*. Experientia 40:339–340

Cole JR, Chai B, Marsh TL, Farris RJ, Wang Q, Kulam SA, Chandra S, McGarrell DM, Schmidt TM, Garrity GM, Tiedje JM (2003) The ribosomal database project (RDP-II): previewing a new autoaligner that allows regular updates and the new prokaryotic taxonomy. Nucl Acids Res 31:442–443

Consortium THMJRS (2010) A catalog of reference genomes from the human microbiome. Science 328:994–999

Conway KR, Boddy CN (2013) ClusterMine360: a database of microbial PKS/NRPS biosynthesis. Nucleic Acids Res 41:D402–D407

Cortina NS, Krug D, Plaza A, Revermann O, Müller R (2012) Myxoprincomide: a natural product from *Myxococcus xanthus* discovered by comprehensive analysis of the secondary metabolome. Angew Chem Int Ed 51:811–816

Cragg GM, Newman DJ (2013) Natural products: a continuing source of novel drug leads. Biochim Biophys Acta 1830:3670–3695

Craney A, Ozimok C, Pimentel-Elardo SM, Capretta A, Nodwell JR (2012) Chemical perturbation of secondary metabolism demonstrates important links to primary metabolism. Chem Biol 19:1020–1027

Cueto M, Jensen PR, Kauffman C, Fenical W, Lobkovsky E, Clardy J (2001) Pestalone, a new antibiotic produced by a marine fungus in response to bacterial challenge. J Nat Prod 64:1444–1446

Cunningham KG, Manson W, Spring FS (1950) Cordycepin, a metabolic product isolated from cultures of *Cordyceps militaris* (Linn.) Link. Nature 166:949

Degnan BM, Hawkins CJ, Lavin MF, McCaffrey EJ, Parry DL, Van den Brenk AL, Watters DJ (1989) New cyclic peptides with cytotoxic activity from the ascidian *Lissoclinum patella*. J Med Chem 32:1349–1354

Deng C, Pan B, O'Connor OA (2013) Brentuximab vedotin. Clin Can Res 19:22–27

Densmore D (2009) A platform-based design environment for synthetic biological systems. TAPIA '09 The fifth Richard Tapia celebration of diversity in computing conference: intellect, initiatives, insight, and innovations. ACM, New York

DeSantis TZ, Hugenholtz P, Larsen N, Rojas M, Brodie EL, Keller K, Huber T, Dalevi D, Hu P, Andersen GL (2006) Greengenes, a chimera-checked 16S rRNA gene database and workbench compatible with ARB. App Environ Microbiol 72:5069–5072

Donia MS, Fricke WF, Partensky F, Cox J, Elshahawi SI, White JR, Phillippy AM, Schatz MC, Piel J, Haygood MG, Ravel J, Schmidt EW (2011) Complex microbiome underlying secondary and primary metabolism in the tunicate-*Prochloron* symbiosis. Proc Nat Acad Sci USA 108:E1423–E1432

Donia MS, Ruffner DE, Cao S, Schmidt EW (2011) Accessing the hidden majority of marine natural products through metagenomics. ChemBioChem 12:1230–1236

Doroghazi JR, Albright JC, Goering AW, Ju K-S, Haines RR, Tchalukov KA, Labeda DP, Kelleher NL, Metcalf WW (2014) A roadmap for natural product discovery based on large-scale genomics and metabolomics. Nat Chem Biol 10:963–968

Ebada SS, Fischer T, Klaβen S, Hamacher A, Roth YO, Kassack MU, Roth EH (2014) A new cytotoxic steroid from co-fermentation of two marine alga-derived micro-organisms. Nat Prod Res 28(16):1–5

Eldridge GR, Vervoort HC, Lee CM, Cremin PA, Williams CT, Hart SM, Goering MG, O'Neil-Johnso M, Zeng L (2002) High-throughput method for the production and analysis of large natural product libraries for drug discovery. Anal Chem 74:3963–3971

Ellis T, Adie T, Baldwin GS (2011) DNA assembly for synthetic biology: from parts to pathways and beyond. Integr Biol 3:109–118

Engler C, Marillonnet S (2011) Generation of families of construct variants using golden gate shuffling. In: Lu C, Browse J, Wallis JG (eds) cDNA Libraries, vol 729. Methods in molecular biology. Humana Press, Totowa, pp 167–181

Eustáquio AS, Moore BS (2008) Mutasynthesis of fluorosalinosporamide, a potent and reversible inhibitor of the proteasome. Angew Chem Int Ed 47:3936–3938

Eustáquio AS, O'Hagan D, Moore BS (2010) Engineering fluorometabolite production: fluorinase expression in *Salinispora tropica* yields fluorosalinosporamide. J Nat Prod 73:378–382

Feling RH, Buchanan GO, Mincer TJ, Kauffman CA, Jensen PR, Fenical W (2003) Salinosporamide A: a highly toxic proteasome inhibitor deom a novel marine source, a marine bacterium of the new genus *Salinispora*. Angew Chem Int Ed 42:355–357

Feng Z, Kim JH, Brady SF (2010) Fluostatins produced by the heterologous expression of a TAR reassembled environmental DNA derived type II PKS gene cluster. J Am Chem Soc 132:11902–11903

Flynn EH, Sigal MV Jr, Wiley PF, Gerzon K (1954) Erythromycin I. Properties and degradation studies. J Am Chem Soc 76:3121–3131

Forner D, Berrué F, Correa H, Duncan K, Kerr RG (2013) Chemical dereplication of marine actinomycetes by liquid chromatography-high resolution mass spectrometry profiling and statistical analysis. Anal Chim Acta 805:70–79

Gao X, Wang P, Tang Y (2010) Engineered polyketide biosynthesis and biocatalysis in *Escherichia coli*. Appl Microbiol Biotechnol 88:1233–1242

Gerth K, Irschik H, Reichenbach H, Trowitzsch W (1980) Myxothiazol, an antibiotic from *Myxococcus fulvus* (Myxobacterales) I. Cultivation, isolation, physico-chemical and biological properties. J Antibiot 33:1474–1479

Gibson DG, Young L, Chuang R-Y, Venter JC, Hutchison CA, Smith HO (2009) Enzymatic assembly of DNA molecules up to several hundred kilobases. Nat Meth 6:343–345

Giddings L-A, Newman D (2013) Microbial natural products: molecular blueprints for antitumor drugs. J Ind Microbiol Biotechnol 40:1181–1210

Gil RR (2011) Constitutional, configurational, and conformational analysis of small organic molecules on the basis of NMR residual dipolar couplings. Angew Chem Int Ed 50:7222–7224

Gomez-Escribano J, Bibb M (2014) Heterologous expression of natural product biosynthetic gene clusters in Streptomyces coelicolor: from genome mining to manipulation of biosynthetic pathways. J Ind Micro Biotech 41:425–431

Gomez-Escribano JP, Song L, Fox DJ, Yeo V, Bibb MJ, Challis GL (2012) Structure and biosynthesis of the unusual polyketide alkaloid coelimycin P1, a metabolic product of the cpk gene cluster of Streptomyces coelicolor M145. Chem Sci 3:2716–2720

Grindberg RV, Ishoey T, Brinza D, Esquenazi E, Coates RC, Liu W-t, Gerwick L, Dorrestein PC, Pevzner P, Lasken R (2011) Single cell genome amplification accelerates identification of the apratoxin biosynthetic pathway from a complex microbial assemblage. PLoS ONE 6:e18565

Halabalaki M, Vougogiannopoulou K, Mikros E, Skaltsounis AL (2014) Recent advances and new strategies in the NMR-based identification of natural products. Curr Opin Biotech 25:1–7

Handelsman J, Rondon MR, Brady SF, Clardy J, Goodman RM (1998) Molecular biological access to the chemistry of unknown soil microbes: a new frontier for natural products. Chem Biol 5:R245–R249

He H, Ding W-D, Bernan VS, Richardson AD, Ireland CM, Greenstein M, Ellestad GA, Carter GT (2001) 123:5362–5363

He R, Wakimoto T, Takeshige Y, Egami Y, Kenmoku H, Ito T, Wang B, Asakawa Y, Abe I (2012) Porphyrins from a metagenomic library of the marine sponge *Discodermia calyx*. Mol BioSyst 8:2334–2338

Hildebrand M, Waggoner LE, Liu H, Sudek S, Allen S, Anderson C, Sherman DH, Haygood M (2004) bryA: an unusual modularpolyketide synthase gene from the uncultivated bacterial symbiont of the marine bryozoan *Bugula neritina*. Chem Biol 11:1543–1552

Ho T-F, Ma C-J, Lu C-H, Tsai Y-T, Wei Y-H, Chang J-S, Lai J-K, Cheuh P-J, Yeh C-T, Tang P-C, Tsai Chang J, Ko J-L, Liu F-S, Yen HE, Chang C-C (2007) Undecylprodigiosin selectively induces apoptosis in human breast carcinoma cells independent of p53. Toxicol Appl Pharmacol 225:318–328

Hosaka T, Ohnishi-Kameyama M, Muramatsu H, Murakami K, Tsurumi Y, Kodani S, Yoshida M, Fujie A, Ochi K (2009) Antibacterial discovery in actinomycetes strains with mutations in RNA polymerase or ribosomal protein S12. Nat Biotech 27:462–464

Huson D, Auch A, Qi J, Schuster S (2007) MEGAN analysis of metagenomic data. Genome Res 17:377–386

Ichikawa N, Sasagawa M, Yamamoto M, Komaki H, Yoshida Y, Yamazaki S, Fujita N (2013) DoBISCUIT: a database of secondary metabolite biosynthetic gene clusters. Nucleic Acids Res 41:D408–D414

Igarashi Y, Kim Y, In Y, Ishida T, Kan Y, Fujita T, Iwashita T, Tabata H, Onaka H, Furumai T (2010) Alchivemycin A, a bioactive polycyclic polyketide with an unprecedented skeleton from *Streptomyces* sp. Org Lett 12:3402–3405

Imada C, Okami Y, Hatta K (2002) Production of selenohomocystine as an antibiotic by a marine *Bacillus* sp. no. 14 with selenomethionine resistance. J Antibiot 55:223–226

Ishida K, Lincke T, Behnken S, Hertweck C (2010) Induced biosynthesis of cryptic polyketide metabolites in a *Burkholderia thailandensis* quorum sensing mutant. J Am Chem Soc 132:13966–13968

Itaya M, Fujita K, Kuroki A, Tsuge K (2008) Bottom-up genome assembly using the *Bacillus subtilis* genome vector. Nat Meth 5:41–43

Itoh N, Kariya S, Kurokawa J (2014) Efficient PCR-based amplification of diverse alcohol dehydrogenase genes from metagenomes for improving biocatalysis: screening of gene-specific amplicons from metagenomes. App Environ Microbiol 80:6280–6289

Iwai S, Chai B, Sul WJ, Cole JR, Hashsham SA, Tiedje JM (2009) Gene-targeted-metagenomics reveals extensive diversity of aromatic dioxygenase genes in the environment. ISME J 4:279–285

Iwasaki S, Kobayashi H, Furukawa J, Namikoshi M, Okuda S, Sato Z, Matsuda I, Noda T (1984) Studies on macrocyclic lactone antibiotics. VII. Structure of a phytotoxin "rhizoxin" produced by *Rhizopus chinensis*. J Antibiot 37:354–362

Jia B, Xuan L, Cai K, Hu Z, Ma L, Wei C (2013) NeSSM: a next-generation sequencing simulator for metagenomics. PLoS ONE

Johnson TA, Sohn J, Inman WD, Estee SA, Loveridge ST, Vervoort HC, Tenney K, Liu J, Ang KK-H, Ratnam J, Bray WM, Gassner NC, Shen YY, Lokey RS, McKerrow JH, Boundy-Mills K, Nukanto A, Kanti A, Julistiono H, Kardono LBS, Bjeldanes LF, Crews P (2011) Natural product libraries to accelerate the high-throughput discovery of therapeutic leads. J Nat Prod 74:2545–2555

Kaeberlein T, Lewis K, Epstein SS (2002) Isolating "uncultivable" microorganisms in pure culture in a simulated natural environment. Science 296:1127–1129

Kaselein JJ, Maki KC, Susekov A, Exhov M, Nortestgaard BG, Machielse BN, Kling D, Davidson MH (2014) Omega -3 free fatty acids for the treatment of severe hypertriglyceridemia. The EVOLVE trial. J Clin Lipdol 8:94–106

Kawai K, Wang G, Okamoto S, Ochi K (2007) The rare earth, scandium, causes antibiotic overproduction in *Streptomyces* sp. FEMS Microbiol Lett 274:311–315

Kersten RD, Lane AL, Nett M, Richter TKS, Duggan BM, Dorrestein PC, Moore BS (2013a) Bioactivity-guided genome mining reveals the lomaiviticin biosynthetic gene cluster in *Salinispora tropica*. ChemBioChem 14:955–962

Kersten RD, Yang Y-L, Xu Y, Cimermancic P, Nam S-J, Fenical W, Fischbach MA, Moore BS, Dorrestein PC (2011) A mass spectrometry–guided genome mining approach for natural product peptidogenomics. Nat Chem Biol 7:794–802

Kersten RD, Ziemert N, Gonzalez DJ, Duggan BM, Nizet V, Dorrestein PC, Moore BS (2013b) Glycogenomics as a mass spectrometry-guided genome-mining method for microbial glycosylated molecules. Proc Nat Acad Sci USA 110:E4407–E4416

Kim JH, Feng Z, Bauer JD, Kallifidas D, Calle PY, Brady SF (2010) Cloning large natural product gene clusters from the environment: piecing environmental DNA gene clusters back together with TAR. Biopolymers 93:833–844

Kishi Y, Fang FG, Forsyth CJ, Scola PM, Yoon SK (1995) Halichondrins and related compounds. US Patent 5436238 A, 26JUL1995

Kouprina N, Larionov V (2008) Selective isolation of genomic loci from complex genomes by transformation-associated recombination cloning in the yeast *Saccharomyces cerevisiae*. Nat Protocols 3:371–377

Langer-Safer PR, Levine M, Ward DC (1982) Immunological method for mapping genes on *Drosophila* polytene chromosomes. Proc Nat Acad Sci USA 79:4381–4385

Larionov V, Kouprina N, Graves J, Chen XN, Korenberg JR, Resnick MA (1996) Specific cloning of human DNA as yeast artificial chromosomes by transformation-associated recombination. Proc Nat Acad Sci USA 93:491–496

Laureti L, Song L, Huang S, Corre C, Leblond P, Challis GL, Aigle B (2011) Identification of a bioactive 51-membered macrolide complex by activation of a silent polyketide synthase in *Streptomyces ambofaciens*. Proc Nat Acad Sci USA 108:6258–6263

Lewis K, Moore C, Zhang Q, Peoples A (2009) Novel macrocyclic polyene lactams. US2009149395, 10DEC2008

Lewis K, Rothfeder M, Ling LL, Moore C, Zhang Q, Peoples A (2012) Macrolactam compounds. US 8097709 B2, 17JAN2012

Li MZ, Elledge SJ (2007) Harnessing homologous recombination in vitro to generate recombinant DNA via SLIC. Nat Meth 4:251–256

Li MZ, Elledge SJ (2012) SLIC: a method for sequence- and ligation-independent cloning. In: Peccoud J (ed) Gene synthesis, vol 852. Methods in Molecular Biology. Humana Press, Totowa, pp 51–59

Lim FY, Sanchez JF, Wang CCC, Keller NP (2012) Chapter fifteen—toward awakening cryptic secondary metabolite gene clusters in filamentous fungi. In: David AH (ed) Methods in enzymology, vol 517. Academic Press, New York, pp 303–324

Long PF, Dunlap WC, Battershill CN, Jaspars M (2005) Shotgun cloning and heterologous expression of the patellamide gene cluster as a strategy to achieving sustained metabolite production. ChemBioChem 6:1760–1765

Luesch H, Yoshida WY, Moore RE, Paul VJ, Corbett TH (2001) Total structure determination of apratoxin A, a potent novel cytotoxin from the marine cyanobacterium *Lyngbya majuscula*. J Am Chem Soc 123:5418–5423

Marmann A, Aly AH, Lin W, Wang B, Proksch P (2014) Co-cultivation—a powerful emerging tool for enhancing the chemical diversity of microorganisms. Mar Drugs 12:1043–1065

Marques JV, Kim K-W, Lee C, Costa MA, May GD, Crow JA, Davin LB, Lewis NG (2013) Next generation sequencing in predicting gene function in podophyllotoxin biosynthesis. J Biol Chem 288:466–479

Martin JR, Rosenbrook W (1967) Studies on the biosynthesis of the erythromycins. II. Isolation and structure of a biosynthetic intermediate, 6-deoxyerythronolide B. Biochemistry 6:435–440

Masuda M, Urabe E, Sakurai A, Sakakibara M (2006) Production of cordycepin by surface culture using the medicinal mushroom *Cordyceps militaris*. Enzyme Microb Tech 39:641–646

McGuire JM, Bunch RL, Anderson RC, Boaz HE, Flynn EH, Powell HM, Smith JW (1952) Ilotycin, a new antibiotic. Antibiot Chemother 2:281–283

Milshteyn A, Schneider JS, Brady SF (2014) Mining the metabiome: identifying novel natural products from microbial communities. Chem Biol 21:1211–1223

Moore JM, Bradshaw E, Seipke RF, Hutchings MI, McArthur M (2012) Chapter eighteen—use and discovery of chemical elicitors that stimulate biosynthetic gene clusters in *Streptomyces* bacteria. In: Hopwood DA (ed) Methods in enzymology, vol 517. Academic Press, London, pp 367–385

Müller K, Faeh C, Diederich F (2007) Fluorine in pharmaceuticals: looking beyond intuition. Science 317:1881–1886

Nenkep VN, Yun K, Li Y, Choi HD, Kang JS, Son BW (2010) New production of haloquinones, bromochlorogentisylquinones A and B, by a halide salt from a marine isolate of the fungus *Phoma herbarum*. J Antibiot 63:199–201

Nett M, Gulder TAM, Kale AJ, Hughes CC, Moore BS (2009) Function-oriented biosynthesis of β-lactone proteasome inhibitors in *Salinispora tropica*. J Med Chem 52:6163–6167

Newman DJ (2012) The Bryostatins. In: Cragg GM, Kingstonn DGI, Newman DJ (Eds) Anticancer agents from natural products, 2nd edn. CRC Press, Boca Raton, pp 199–218

Newman DJ, Cragg GM (2012) Natural products as sources of new drugs over the 30 years from 1981 to 2010. J Nat Prod 75:311–335

Ochi K (2007) From microbial differentiation to Ribosome Engineering. Biosci Biotechnol Biochem 71:1373–1386

Ochi K, Hosaka T (2013) New strategies for drug discovery: activation of silent or weakly expressed microbial gene clusters. Appl Microbiol Biotechnol 97:87–98

Ochi K, Tanaka Y, Tojo S (2014) Activating the expression of bacterial cryptic genes by rpoB mutations in RNA polymerase or by rare earth elements. J Ind Microbiol Biotechnol 41:403–414

Oh D-C, Jensen PR, Kauffman CA, Fenical W (2005) Libertellenones A-D: induction of cytotoxic diterpenoid biosynthesis by marine microbial competition. Bioorg Med Chem 13:5267–5273

Oh D-C, Kauffman CA, Jensen PR, Fenical W (2007) Induced production of emericellamides A and B from the marine-derived fungus *Emericella* sp. in competing co-culture. J Nat Prod 70:515–520

Oh D-C, Poulsen M, Currie CR, Clardy J (2009) Dentigerumycin: a bacterial mediator of an ant-fungus symbiosis. Nat Chem Biol 5:391–393

Ohnishi Y, Ishikawa J, Hara H, Suzuki H, Ikenoya M, Ikeda H, Yamashita A, Hattori M, Horinouchi S (2008) Genome sequence of the streptomycin-producing microorganism *Streptomyces griseus* IFO 13350. J Bacteriol 190:4050–4060

Onaka H, Mori Y, Igarashi Y, Furumai T (2011) Mycolic acid-containing bacteria induce natural-product biosynthesis in *Streptomyces* species. App Environ Microbiol 77:400–406

Ongley SE, Bian X, Neilan BA, Müller R (2013) Recent advances in the heterologous expression of microbial natural product biosynthetic pathways. Nat Prod Rep 30:1121–1138

Olivera BM, Cruz LJ, deSantos V, LeCheminant GW, Griffin D, Zeikus R, McIntosh JM, Galyean R, Varga J (1987) Neuronal calcium channel antagonists. Discrimination between calcium channel subtypes using ω-conotoxin from *Conus magus* venom. Biochemistry 26:2085–2090

Owen JG, Reddy BVB, Ternei MA, Charlop-Powers Z, Calle PY, Kim JH, Brady SF (2013) Mapping gene clusters within arrayed metagenomic libraries to expand the structural diversity of biomedically relevant natural products. Proc Nat Acad Sci USA 110:11797–11802

Park HB, Kwon HC, Lee C-H, Yang HO (2009) Glionitrin A, an antibiotic-antitumor metabolite derived from competitive interaction between abandoned mine microbes. J Nat Prod 72:248–252

Park HB, Kim Y-J, Park J-S, Yang HO, Lee KR, Kwon HC (2011) Glionitrin B, a cancer invasion inhibitory diketopiperazine produced by microbial coculture. J Nat Prod 74:2309–2312

Partida-Martinez LP, Hertweck C (2005) Pathogenic fungus harbours endosymbiotic bacteria for toxin production. Nature 437:884–888

Partida-Martinez LP, Hertweck C (2007) A gene cluster encoding rhizoxin biosynthesis in *Burkholderia rhizoxina*, the bacterial endosymbiont of the fungus *Rhizopus microsporus*. ChemBioChem 8:41–45

Penesyan A, Ballestriero F, Daim M, Kjelleberg S, Thomas T, Egan S (2013) Assessing the effectiveness of functional genetic screens for the identification of bioactive metabolites. Mar Drugs 11:40–49

Pettit RK (2011) Small-molecule elicitation of microbial secondary metabolites. Microb Biotechnol 4:471–478

Pfeifer B, Hu Z, Licari P, Khosla C (2002) Process and metabolic strategies for improved production of *Escherichia coli*-derived 6-deoxyerythronolide B. App Environ Microbiol 68:3287–3292

Pfeifer BA, Admiraal SJ, Gramajo H, Cane DE, Khosla C (2001) Biosynthesis of complex polyketides in a metabolically engineered strain of *E. coli*. Science 291:1790–1792

Piel J (2011) Approaches to capturing and designing biologically small molecules produced by uncultured microbes. Ann Rev Microbiol 65:431–453

Pruesse E, Quast C, Knittel K, Fuchs BM, Ludwig W, Peplies J, Glöckner FO (2007) SILVA: a comprehensive online resource for quality checked and aligned ribosomal RNA sequence data compatible with ARB. Nucl Acids Res 35:7188–7196

Quan J, Tian J (2011) Circular polymerase extension cloning for high-throughput cloning of complex and combinatorial DNA libraries. Nat Protocols 6:242–251

Quan J, Tian J (2014) Circular polymerase extension cloning. In: Valla S, Lale R (eds) DNA cloning and assembly methods, vol 1116. Methods in molecular biology. Humana Press, Totowa, pp 103–117

Rath CM, Janto B, Earl J, Ahmed A, HU FZ, Miller L, Dahlgren M, Kreft R, Yu F, Wolff JJ, Kweon HK, Christiansen MA, Hakansson K, Williams RM, Ehrlich GD, Sherman DH (2011) Meta-omic characterization of the marine invertebrate microbial consortium that produces the chemotherapeutic natural product ET-743. ACS Chem Biol 6:1244–1256

Reddy BVB, Milshteyn A, Charlop-Powers Z, Brady SF (2014) eSNaPD: A versatile, web-based bioinformatics platform for surveying and mining natural product biosynthetic diversity from metagenomes. Chem Biol 21:1023–1033

Rigali S, Titgemeyer F, Barends S, Mulder S, Thomae AW, Hopwood DA, van Wezel GP (2008) Feast or famine: the global regulator DasR links nutrient stress to antibiotic production by *Streptomyces*. EMBO Rep 9:670–675

Rivero-Müller A, Lajić S, Huhtaniemi I (2007) Assisted large fragment insertion by Red/ET-recombination (ALFIRE)–an alternative and enhanced method for large fragment recombineering. Nucleic Acids Res 35:e78

Rodriguez E, Menzella HG, Gramajo H (2009) Heterologous production of polyketides in bacteria. Methods Enzymol 459:339–365

Ross AC, Gulland LES, Dorrestein PC, Moore BS (2014) Targeted capture and heterologous expression of the *Pseudoalteromonas* alterochromide gene cluster in *Escherichia coli* represents a promising natural product exploratory platform. ACS Synth Biol. doi:10.1021/sb500280q

Sakemi S, Ichiba T, Kohmoto S, Saucy G, Higa T (1988) Isolation and structure determination of onnamide A, a new bioactive metabolite of a marine sponge, *Theonella* sp. J Am Chem Soc 110:4851–4853

Sanchez JF, Somoza AD, Keller NP, Wang CCC (2012) Advances in *Aspergillus* secondary metabolite research in the post-genomic era. Nat Prod Rep 29:351–371

Sandmann A, Frank B, Müller R (2008) A transposon-based strategy to scale up myxothiazol production in myxobacterial cell factories. J Bacter 135:255–261

Schroeckh V, Nützmann H-W, Brakhage AA (2014) Fungal-actinomycete interactions–wakening of silent fungal secondary metabolism gene clusters via interorganismic interactions. In: Osbourn A, Goss R, Carter GT (eds) Natural products: discourse, diversity, and design. Wiley, Oxford, pp 147–158

Schroeckh V, Scherlach K, Nützmann H-W, Shelest E, Schmidt-Heck W, Schuemann J, Martin K, Hertweck C, Brakhage AA (2009) Intimate bacterial-fungal interaction triggers biosynthesis of archetypal polyketides in *Aspergillus nidulans*. Proc Nat Acad Sci USA 106:14558–14563

Schulze CJ, Bray WM, Woerhmann MH, Stuart J, Lokey RS, Linington RG (2013) "Function-first" lead discovery: mode of action profiling of natural product libraries using image-based screening. Chem Biol 20:285–295

Schulze-Sünninghausen D, Becker J, Luy B (2014) Rapid heteronuclear single quantum correlation NMR spectra at natural abundance. J Am Chem Soc 136:1242–1245

Scott JJ, Oh D-C, Yuceer MC, Klepzig KD, Clardy J, Currie CR (2008) Bacterial protection of beetle-fungus mutualism. Science 322:63

Senior MM, Williamson RT, Martin GE (2013) Using HMBC and ADEQUATE NMR data to define and differentiate long-range coupling pathways: is the Crews rule obsolete? J Nat Prod 76:2088–2093

Seshime Y, Juvvadi PR, Kitamoto K, Ebizuka Y, Fujii I (2010) Identification of cspyrone B1 as the novel product of *Aspergillus oryzae* type III polyketide synthase *CsyB*. Bioorg Med Chem 18:4542–4546

Shendure J, Ji H (2008) Next-generation DNA sequencing. Nat. Biotech 26:1135–1145

Shetty RP, Endy D, Knight TF Jr (2008) Engineering BioBrick vectors from BioBrick parts. J Biol Eng 2:1–12

Shih C-J, Chen P-Y, Liaw C-C, Lai Y-M, Yang Y-L (2014) Bringing microbial interactions to light using imaging mass spectrometry. Nat Prod Rep 31:739–755

Shima J, Hesketh A, Okamoto S, Kawamoto S, Ochi K (1996) Induction of actinorhodin production by rpsL (encoding ribosomal protein S12) mutations that confer streptomycin resistance in Streptomyces lividans and Streptomyces coelicolor A3(2). J Bacter 178:7276–7284

Siegl A, Kamke J, Hochmuth T, Piel J, Richter M, Liang C, Dandekar T, Hentschel U (2011) Single-cell genomics reveals the lifestyle of *Poribacteria*, a candidate phylum symbiotically associated with marine sponges. ISME J 5:61–70

Simon C, Daniel R (2009) Achievements and new knowledge unraveled by metagenomic approaches. Appl Microbiol Biotechnol 85:265–276

Speitling M, Smetanina OF, Kuznetsova TA, Laatsch H (2007) Bromoalterochromides A and A′, unprecedented chromopeptides from a marine *Pseudoalteromonas maricaloris* strain KMM 636^T. J Antibiot 60:36–42

Stankovic N, Radulovic V, Petkovic M, Vuckovic I, Jadranin M, Vasiljevic B, Nikodinovic-Runic J (2012) *Streptomyces* sp. JS520 produces exceptionally high quantities of undecylprodigiosin with antibacterial, antioxidative, and UV-protective properties. Appl Microbiol Biotech 96:1217–1231

Stepanauskas R (2012) Single cell genomics: an individual look at microbes. Curr Opin Microbiol 15:613–620

Sudek S, Lopanik NB, Waggoner LE, Hildebrand M, Anderson C, Liu H, Patel A, Sherman DH, Haygood MG (2006) Identification of the putative bryostatin polyketide synthase gene cluster from "Candidatus Endobugula ser tula", the uncultivated microbial symbiont of the marine bryozoan Bugula neritina. J Nat Prod 70:67–74

Suckling CJ (1991) Chemical approaches to the discovery of new drugs. Sci Prog Edin 75:323–360

Talá A, Wang G, Zemanova M, Okamoto S, Ochi K, Alifano P (2009) Activation of dormant bacterial genes by *Nonomuraea* sp. strain ATCC 39727 mutant-type RNA polymerase. J Bacteriol 191:805–814

Tanaka Y, Hosaka T, Ochi K (2010) Rare earth elements activate the secondary metabolite-biosynthetic gene clusters in *Streptomyces coelicolor* A3(2). J Antibiot 63:477–481

Torsvik V, Daae FL, Sandaa R-A, Øvreås L (1998) Novel techniques for analysing microbial diversity in natural and perturbed environments. J Biotech 64:53–62

Torsvik V, Goksøyr J, Daae FL (1990) High diversity in DNA of soil bacteria. App Environ Microbiol 56:782–787

Tuli HS, Sandhu SS, Shama AK (2014) Pharmacological and therapeutic potential of *Cordyceps* with special reference to cordycepin. 3. Biotech 4:1–12

Tyson GW, Chapman J, Hugenholtz P, Allen EE, Ram RJ, Richardson PM, Solovyev VV, Rubin EM, Rokhsar DS, Banfield JF (2004) Community structure and metabolism through reconstruction of microbial genomes from the environment. Nature 428:37–43

Udwary DW, Zeigler L, Asolkar RN, Singan V, Lapidus A, Fenical W, Jensen PR, Moore BS (2007) Genome sequencing reveals complex secondary metabolome in the marine actinomycete *Salinispora tropica*. Proc Nat Acad Sci USA 104:10376–10381

Venter JC, Remington K, Heidelberg JF, Halpern AL, Rusch D, Eisen JA, Wu D, Paulsen I, Nelson KE, Nelson W, Fouts DE, Levy S, Knap AH, Lomas MW, Nealson K, White O, Peterson J, Hoffman J, Parsons R, Baden-Tillson H, Pfannkoch C, Rogers Y-H, Smith HO (2004) Environmental genome shotgun sequencing of the Sargasso Sea. Science 304:66–74

von Döhren H (2009) A survey of nonribosomal peptide synthetase (NRPS) genes in *Aspergillus nidulans*. Fungal Genetics and Biology 46:S45–S52

von Mering C, Hugenholtz P, Raes J, Tringe SG, Doerks T, Jensen LJ, Ward N, Bork P (2007) Quantitative phylogenetic assessment of microbial communities in diverse environments. Science 315:1126–1130

Wang CCC, Chiang Y-M, Praseuth MB, Kuo P-L, Liang H-L, Hsu Y-L (2010) Asperfuranone from *Aspergillus nidulans* inhibits proliferation of human non-small cell lung cancer A549 cells via blocking cell cycle progression and inducing apoptosis. Basic Clin Pharmacol Toxicol 107:583–589

Watrous JD, Dorrestein PC (2011) Imaging mass spectrometry in microbiology. Nat Rev Micro 9:683–694

Wiley PF, Gerzon K, Flynn EH, Sigal MV, Weaver O, Quarck UC, Chauvette RR, Monahan R (1957) Erythromycin. X.1 Structure of erythromycin. J Am Chem Soc 79:6062–6070

Williamson RT, Buevich AV, Martin GE (2014) Using LR-HSQMBC to observe long-range $^1H–^{15}N$ correlations. Tetrahedron Lett 55:3365–3366

Wilson MC, Piel J (2013) Metagenomic approaches for exploiting uncultivated bacteria as a resource for novel biosynthetic enzymology. Chem Biol 20:636–647

Wilson MC, Mori T, Rückert C, Uria AR, Helf MJ, Takada K, Gernert C, Steffens UAE, Heycke N, Schmitt S, Rinke C, Helfrich EJN, Brachmann AO, Gurgui C, Wakimoto T, Kracht M, Crüsemann M, Hentschel U, Abe I, Matsunaga S, Kalinowski J, Takeyama H, Piel J (2014) An environmental bacterial taxon with a large and distinct metabolic repertoire. Nature 506:58–62

Wings S, Müller H, Berg G, Lamshöft M, Leistner E (2013) A study of the bacterial community in the root system of the maytansine containing plant *Putterlickia verrucosa*. Phytochemistry 91:158–164

Winnikoff JR, Glukhov E, Watrous J, Dorrestein PC, Gerwick WH (2014) Quantitative molecular networking to profile marine cyanobacterial metabolomes. J Antibiot 67:105–112

Woo CM, Beizer NE, Janso JE, Herzon SB (2012) Isolation of lomaiviticins C-E, transformation of lomaiviticin C to lomaiviticin A, complete structure elucidation of lomaiviticin A, and structure—activity analyses. J Am Chem Soc 134:15285–15288

Wood D, Salzberg S (2014) Kraken: ultrafast metagenomic sequence classification using exact alignments. Genome Biol 15:R46

Wright LF, Hopwood DA (1976) Actinorhodin is a chromosomally-determined antibiotic in *Streptomyces coelicolor A3(2)*. J Gen Microbiol 96:289–297

Wu D, Hugenholtz P, Mavromatis K, Pukall R, Dalin E, Ivanova NN, Kunin V, Goodwin L, Wu M, Tindall BJ, Hooper SD, Pati A, Lykidis A, Spring S, Anderson IJ, D'haeseleer P, Zemla A, Singer M, Lapidus A, Nolan M, Copeland A, Han C, Chen F, Cheng J-F, Lucas S, Kerfeld C, Lang E, Gronow S, Chain P, Bruce D, Rubin EM, Kyrpides NC, Klenk H-P, Eisen JA (2009) A phylogeny-driven genomic encyclopaedia of *Bacteria* and *Archaea*. Nature 462:1056–1060

Xu R, Greiveldinger G, Marenus LE, Cooper A, Ellman JA (1999) Combinatorial library approach for the identification of synthetic receptors targeting vancomycin-resistant bacteria. J Am Chem Soc 121:4898–4899

Yamanaka K, Reynolds KA, Kersten RD, Ryan KS, Gonzalez DJ, Nizet V, Dorrestein PC, Moore BS (2014) Direct cloning and refactoring of a silent lipopeptide biosynthetic gene cluster yields the antibiotic taromycin A. Proc Nat Acad Sci USA 111:1957–1962

Yang Y-L, Xu Y, Kersten RD, Liu W-T, Meehan MJ, Moore BS, Bandeira N, Dorrestein PC (2011) Connecting chemotypes and phenotypes of cultured marine microbial assemblages by imaging mass spectrometry. Angew Chem Int Ed 50:5839–5842

Yarnold JE, Hamilton BR, Welsh DT, Pool GF, Venter DJ, Carroll AR (2012) High resolution spatial mapping of brominated pyrrole-2-aminoimidazole alkaloids distributions in the marine sponge *Stylissa flabellata* via MALDI-mass spectrometry imaging. Mol BioSyst 8:2249–2259

Yeh H-H, Chiang Y-M, Entwistle R, Ahuja M, Lee K-H, Bruno K, Wu T-K, Oakley B, Wang CC (2012) Molecular genetic analysis reveals that a nonribosomal peptide synthetase-like (NRPS-like) gene in *Aspergillus nidulans* is responsible for microperfuranone biosynthesis. Appl Microbiol Biotechnol 96:739–748

Youssef NH, Couger MB, McCully AL, Criado AEG, Elshahed MS (2014) Assessing the global phylum level diversity within the bacterial domain: a review. J Adv Res. doi:10.1016/j.jare.2014.10.005

Zengler K, Toledo G, Rappé M, Elkins J, Mathur EJ, Short JM, Keller M (2002) Cultivating the uncultured. Proc Nat Acad Sci USA 99:15681–15686

Zhang Y, Werling U, Edelmann W (2012) SLiCE: a novel bacterial cell extract-based DNA cloning method. Nucleic Acids Res 40:e55

Zhou Q, Su X, Jing G, Ning K (2014) Meta-QC-Chain: comprehensive and fast quality control method for metagenomic data. Genomics Proteomics Bioinform 12:52–56

Zhu F, Chen G, Chen X, Huang M, Wan X (2011) Aspergicin, a new antibacterial alkaloid produced by mixed fermentation of two marine-derived mangrove epiphytic fungi. Chem Nat Compd 47:767–769

Zhu F, Chen G, Wu J, Pan J (2013) Structure revision and cytotoxic activity of marinamide and its methyl ester, novel alkaloids produced by co-cultures of two marine-derived mangrove endophytic fungi. Nat Prod Res 27:1960–1964

Zhu F, Lin Y (2006) Marinamide, a novel alkaloid and its methyl ester produced by the application of mixed fermentation technique to two mangrove endophytic fungi from the South China Sea. Chin Sci Bull 51:1426–1430

Ziemert N, Podell S, Penn K, Badger JH, Allen E, Jensen PR (2012) The natural product domain seeker NaPDoS: a phylogeny based bioinformatic tool to classify secondary metabolite gene diversity PLoS ONE

Zotchev SB, Sekurova ON, Katz L (2012) Genome-based bioprospecting of microbes for new therapeutics. Curr Opin Biotech 23:941–947